Practical ITAM:

The essential guide for IT Asset Managers

Practical ITAM:

The essential guide for IT Asset Managers

Getting started and making a difference in the field of IT Asset Management.

By Martin Thompson

Published by The ITAM Review, Summer 2017
www.itassetmanagement.net

You must not publish, sell, distribute or reproduce this book or any part of the text without the permission of the author.

Throughout this book the author refers to brands, products and trademark names. Rather than place a trademark symbol at every occurrence, we hereby state that we are using the names only in an editorial fashion with no intention of infringement of the trademark.

This publication is Copyright © 2017 The ITAM Review / Martin Thompson / Enterprise Opinions Limited, all right reserved worldwide.

Published by The ITAM Review.

Enterprise Opinions Limited trading as "The ITAM Review" Care of Basepoint Business Centre, Rivermead Drive, Westlea, Swindon, Wiltshire. SN5 7EX.

www.itassetmanagement.net

ISBN 978-1547011216

Contents

Introduction:	1
Why do we need asset managers?	5
The 12 Box Model	10
SECTION 1: TAXI	19
Authority	21
Plan	36
Team	47
Stakeholders	58
SECTION 2: TAKE-OFF	67
Entitlement	69
Inventory	81
Reporting	96
SECTION 3: CRUISING ALTITUDE	105
Transition	107
Request	115
Dependencies	125
Reclaim	135
Verification	148
What next?	159
The Author	160
The ITAM Review Newsletter	162

Foreword

IT asset management (ITAM) is a young profession. Physical asset management as a profession has been around for much longer—it has a large number of national and international associations and an extensive training, qualification, and certification infrastructure. ITAM still has a way to go. It was born out of necessity only as IT developed. It is the discipline which tries to bring an improved measure of order and control to the fastest-growing, most-innovative, and most-disruptive area of our lives, both work and personal.

ITAM is an exciting discipline, in part because of its many challenges at both the executive level (where value must be delivered to the organisation) and at the pit face (where attention to detail is needed). Training people to meet these challenges is in itself a challenge.

I come from the old school where people read books, manuals, and contracts cover to cover and then learned on the job—too often through making mistakes. Mistakes are great teachers, but not friendly ones.

This book will help bring those new to the profession up to speed in a way that fits the hectic pace of life today. Its 12 modules are manageable in size, and introduce people to most of the issues they will face in real life. These modules are supported by videos (excellent!), an organisational maturity assessment, a personal certification exam, and the ITAM Review online community. The ITAM Review website is another resource to augment the course.

This course is also a good basis from which to start working towards achieving ISO ITAM, integrated with ISO information security management and ISO service management. Being tightly integrated within the rest of IT is one of our profession's major objectives.

David Bicket
Co-editor of ISO/IEC 19770-1:2017 ITAM Requirements

Co-author of the ITIL SAM/ITAM Guide

Former convener of ISO/IEC JTC1 SC7 WG21 (the ISO ITAM working group)

Acknowledgements

The author expresses his gratitude for help in the creation of the book:

To Jean and Chris.

To Ginny Carter for her invaluable coaching.

To Kylie Fowler for her excellent feedback.

Finally to Mark Richardson and team at the FD Centre, original inspiration for the 12 Box Model.

Introduction:

Making a difference in the field of IT asset management

I was recently trying to describe what I do for a living to one my elderly relatives. I settled on "counting computers" as the simplest way of describing it. Maybe "squeezing best value out of the software and hardware a company buys" would have been more accurate. This might not sound like the most tantalising of subjects—perhaps this is why I'm not inundated with invites to dinner parties! But it's a subject I'm passionate about nonetheless, and I've always wanted to learn more.

IT asset management (ITAM) sits at the intersection between IT, procurement, and finance. It's a governance role that attempts to manage any software or hardware purchased as an asset, so that any risk and cost inherent in the purchase can be managed.

Introducing Stella Hampton

Throughout this book we'll hear from Stella Hampton, an IT asset manager who has recently been hired to revive a flagging ITAM team.

We'll join Stella as she builds a business plan, seeks senior management support, and implements a well-respected and high-value ITAM practice.

We'll see first-hand how Stella uses a methodology called the 12 Box Model to prioritise her workload and focus her team's effort on those tasks that will make the biggest impact.

Why do companies bother with ITAM? Because by managing their assets, organisations save money, reduce the risk of fines and penalties, and can be more competitive and operationally efficient. People choose ITAM as a profession because it's challenging and ever changing, and it allows them to communicate with the entire organisation, giving them a level of influence not possible in many other careers.

This book is a culmination of my conversations, practical experience, and research in the field since 2000. It's an introduction to the world of ITAM that will allow you to build a similar fondness for the subject!

WHY READ THIS BOOK?

If you've just stumbled across the subject of ITAM, you can be forgiven if you feel a little overwhelmed. It's not simply matching up what you own with what you've bought.

Start peeling back the layers of the ITAM onion and it soon gets incredibly complex. You might be feeling the pressure of software audits, or pressure to show cost savings or sharpen up your records. Looking at the sheer number of devices and software titles in your environment can be daunting, and there are only so many hours on the clock.

Over the years I've spoken to hundreds of IT asset managers who feel overwhelmed. These conversations made me realise it was important to write an independent industry guide on how to get started and make a difference. Throughout this book I will break the complex topic of ITAM into bite-size chunks and help you prioritise the steps that will make the most difference.

I distil some of the things I've learned along the way in a format that is modular and easy to consume, understand, and implement. I'm passionate about ITAM, and

I'm also passionate about ITAM's potential to be a strategic role in the future. World-class IT asset managers are indispensable to their organisations; they are lynchpins in decision-making and strategic direction for the whole IT department and have considerable career advancement opportunities.

Read this book and you'll understand what your top priorities are, what should be done first, and how a valuable and lasting ITAM practice can be built for your company. The subject is broken down into 12 manageable chunks, which allow you to identify where your strengths and weaknesses lie—and real-life examples and practical strategies show you how to make a real impact.

RECOMMENDED APPROACH

To make the best use of this book, first familiarise yourself with the key topics and then take our free online maturity assessment to identify the strengths and weaknesses in your company. Next, ask questions in the ITAM Review community to get support from your peers. Finally, verify your learning by taking the 12 Box certification exam.

1. Read the book
2. Take the free maturity assessment
3. Get peer support from the ITAM Review community
4. Get your learning verified with 12 Box certification
   https://www.itassetmanagement.net/practical-itam/

LET'S GET STARTED!

Without further ado, let's get started. The sooner we get going, the sooner you can start making an impact.

Follow the guidelines in this book and you'll have a

clear plan of action—a roadmap for building a world-class ITAM practice. There are also many resources to help you, and a community to rely upon if you need help or have any questions.

Why do we need asset managers?

Those new to the subject of IT asset management (ITAM) may be wondering why we need IT asset managers in the first place. It can't be that complicated—can it? Let's first explore why software licensing is complex. This will give you a flavour of the complexity involved in ITAM and of the critical role that IT asset managers perform.

The Internet pioneer and investor Marc Andreessen said, "Software is eating the world". Software runs on your phone, in your car, at your place of work, and on the satellites orbiting the earth. Software is everywhere.

Software is now part of everything we do and permeates every area of business. Software and the hardware supporting it underpin the transactions and systems for entire companies.

We buy software in its many forms via software licensing agreements. Investing in software is rather like hiring a rental car; you don't actually own the car, you just have a right to use it, and how you use it is dependent on the terms set out by the car-hire company. Somewhere in the small print of your agreement will be a clause stating that you are not allowed to race the car on a track or compete in an off-road rally. Whilst no doubt a lot of fun, this wouldn't be very fair to the car-hire company, and if everyone did it, they'd go out of business.

THE ROLE OF SOFTWARE LICENSING

When you buy software you're paying for the right to use it. You don't actually own it. A licence entitles a person or business to use the software under certain conditions, as dictated by the terms and conditions.

Software licensing allows the software creators to dictate how their product can be used and how their intellectual property is protected. Software publishers want to be paid a fair price for the time and effort they invest in developing and supporting their software. They also want to ensure a fair trade of value, i.e., that the price paid for the software is proportional to the value received.

A small greengrocer may want to use the spreadsheet application Microsoft Excel to calculate produce sold and his profit-and-loss statement. The greengrocer can buy a licence from Microsoft to use the software in his business under certain terms. In contrast, a Fortune 100 oil company with thousands of employees worldwide will be sold a different licence for Microsoft Excel to make sure they are being charged a fair price for the value received.

Licensing models allow the publisher to make these discriminations between types of customer and how their software is used. Licensing allows the publisher to capture maximum revenue for the value being delivered whilst being as flexible as possible to customer requirements. In this example, if Microsoft weren't flexible, someone would be getting a bad deal—either it would be far too expensive for the greengrocer or extraordinarily cheap for the oil company. The goal is to maximise revenue from each customer whilst serving as many customers as possible.

LICENSING AND INNOVATION

The example above illustrates a simple but important concept. The complexity of licensing, software pricing, and software product usage rights stems from software publishers' trying to strike the right balance between offering fair value, maximising revenue, and protecting their intellectual property.

The licensing complexity of trying to serve as many customers as possible is made worse by continual innovations in technology. New gadgets, devices, platforms, and exponentially more powerful computing power are being introduced all the time. Therefore, "fair value" is always in a state of flux.

For an example of the power of innovation, let's return to our oil company. Say the geological data for finding new oilfields and setting the optimal price for a barrel of oil is calculated using software from the database giant Oracle. The Oracle database is installed on a piece of IT hardware called a server.

In the 1960s, Gordon Moore, founder of the chip manufacturer Intel, noticed a strange phenomenon in IT hardware that still holds true to this day. Moore's law states that computing power doubles every 18 months.

So logic suggests that within 18 months, the oil company could replace the server that the Oracle database runs on for roughly the same price and twice the performance. This means a faster database, better performance, and potentially more profit per barrel and more new oilfields to harvest. More bang for the same Oracle database buck. Should Oracle be entitled to a share of this additional value being created by their software?

Similarly, our greengrocer might use an iPad tablet to access his Excel sheet and calculate his profit and loss whilst in the comfort of his armchair at home, thereby increasing his quality of life. Should Microsoft allow him to do that with the same licence, even though he is getting much more value?

These are two simple examples of how new innovation brings into question the balance between value for the customer and revenue and intellectual property protection for the publisher.

LICENSING COMPLEXITY

There is a direct relationship between licensing flexibility and complexity. The more new customers the software publisher wants to target, the more complex the licensing rules become. Similarly, each new market the software publisher carves out with special licence terms leads to more complexity.

There is also a relationship between the growth rate and market share of a software company and licensing complexity. Young, innovative companies with high growth looking for market share are likely to be more relaxed about giving the customer more value. Simple-to-understand licence programs with obvious value allow companies to onboard new customers as quickly as possible. But an older software company with low growth and shrinking market share will likely want to squeeze every last drop of revenue possible from existing customers by building complex licence models.

Finally, software publishers can flex their competitive muscles via their licensing programs. Software publishers wishing to compete in highly competitive markets, out-innovate their competitors, and offer prospective customers deals and offers can do so via their licence programs.

It's worth noting that it's not possible for an IT asset manager to have encyclopaedic knowledge of every single licensing program within their environment (at the time of writing, the Microsoft product terms for volume licensing document is 93 pages!). More valuable is the ability to learn and have a broad understanding of how licensing works and key risks to look for.

INTRODUCING ITAM

IT asset management (ITAM) is the business practice of managing software and hardware as an asset. The software and hardware that a business uses has certain terms, costs and restrictions on its use. ITAM allows a business to manage the costs and value of the assets it uses without too much exposure to risk.

Using our car-hire example from above—if a company leased a good number of company cars, it would make sense for someone in the company to manage them. This person could keep track of costs, legal obligations, lease expiry, which member of staff was assigned each car, whether the driver had a licence and appropriate insurance, and so on. This bit of administrative housekeeping would minimise the risk to the company and ensure the company was making best use of its investment in cars. The same is true of IT; an IT asset manager ensures the company makes best use of IT whilst minimising risk.

The 12 Box Model

BREAKING ITAM INTO MANAGEABLE CHUNKS

In this chapter we will cover a framework called the 12 Box Model.

The model is specifically designed for breaking up the complex subject of ITAM into manageable chunks, so that IT asset managers struggling to get their arms around the subject can break off smaller pieces and prioritise their work. The 12 Box Model originates from IT asset management implementation in the real world and conversations with hundreds of ITAM professionals about their wins and failures.

By the end of this chapter you'll know the 12 main areas of the model, the approach you might consider, and how the model should be used. By the time you've finished reading this book I hope the subject will seem a bit less daunting and you feel empowered to make an impact!

BENCHMARKING ITAM MATURITY

Before digging into the 12 Box Model, it's worth highlighting a couple of other ITAM industry frameworks you might find useful.

It's common for IT departments to benchmark their maturity or perform some form of risk assessment before improving their internal practices, such as ITAM. In the

ITAM field, many third parties can help you assess your maturity, and there are several industry models:

- ☐ Microsoft SAM Optimisation Model – This model classes organisations into four categories (basic, standard, rationalised, and dynamic) based on ten questions. It's a good model, and Microsoft should be congratulated for bringing it into the market to support SAM best practices. However, it is Microsoft-centric and geared towards performing a licensing reconciliation with Microsoft rather than building a world-class ITAM practice.
- ☐ ISO/IEC 19770-1 – Similarly, the international standard ISO 19770 is very good but might be a little out of reach if you're just starting out. We cover ISO/IEC 19770-1 in more detail, including when it might be a good time to look at the standard, in Chapter 12 – Verification.

Many readers coming into the ITAM Review are brand new to the subject and are trying to get their heads around the topic. Some have just been assigned the role for the first time, or they've been asked to look at it, or they're applying for a job in this space, or they simply find the subject interesting. They start peeling back layers of the onion and the sheer enormity and complexity of the subject overwhelms them. I wanted to build a model to help people understand the concept of ITAM, assess their own maturity, and then begin to make progress.

The model is vendor-and-service-provider independent and provides the right level of strategic focus for developing a modern ITAM practice. In our model, there are 12 boxes. If you've "ticked all the boxes", you're doing well and are on the way to developing a world-class ITAM practice.

The focus of the model is getting to audit-ready status. The goal is to be ready for an external or internal audit to show the quality of your practice. It's also based on a continual service improvement (CSI) approach. Throughout this book we'll talk CSI. It's about recognising that you don't do ITAM overnight. You can't stand it up in three months. You might be able to do a true-up in three months, but to build a lasting practice you need to approach this as a progressive work stream to improve over time.

PEOPLE, PROCESS AND TECHNOLOGY

The 12 Box Model is a mixture of people, process, and technology. You can't do ITAM properly with just people, or just processes, or just technology. You need the right blend of all three. The 12 Box Model recognises that blend:

- **People** is about winning the hearts and minds of your team, stakeholders, employees, and leadership, and about keeping the right level of focus.
- **Process** is about putting the checks and balances in place to measure what you're using and what you're buying, embracing the fact that your business is in a constant state of change, and using processes to embrace change.
- **Technology** is about record keeping, automating processes where possible using technology, and using all the intelligence that we create with accurate ITAM data to help the business make the right decisions.

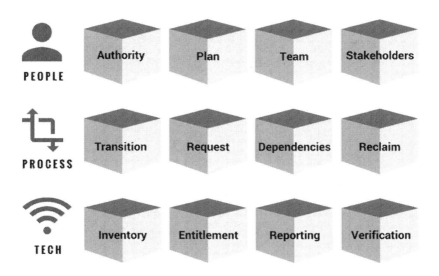

AN OVERVIEW OF THE 12 BOX MODEL

The 12 boxes in the image above represent the 12 focus areas for a world-class ITAM practice.

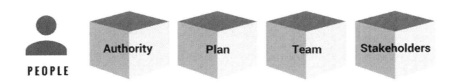

Starting off with the people side of things, we need **authority** first of all. There's no point in proceeding without authority; you might as well go back to bed! You need the blessing of senior management; otherwise, you're not going to have the budget or permission necessary to make change. You need the senior management team to thump the table and ask you to get on with it and remove blockers when they arise.

Then take that authority and build a business **plan**. Plan what you're going to do, when you're going to do it, and what resources you need. Next, recruit a **team** to execute the plan. And because you can't do ITAM in isolation, you also need to work with your **stakeholders** on executing your plan.

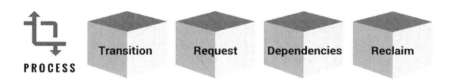

There are over 20 processes in the ISO ITAM standard; we're focused on four key areas. The first one is **transition**, which involves the need to embrace change. Every one of our businesses is in a constant state of transition; if they weren't, then we probably wouldn't have the roles that we have. We don't want to be chasing the business to say, "What have you done? What mess have you made this week?" We need work with the source of change to make sure that our ITAM practices are baked into everyday processes.

We then look at **requests**, which may vary from "I need a copy of Visio" to "I need a new laptop" to "I need to build three new data centres".

Next come our **dependencies**. There are several different data sets within ITAM. We have users, we have devices, we have systems, we have priorities, we have usage, and more. All of these data sets have dependencies between them. If we're going to manage our risk and our costs, we need a process for managing dependencies.

If request means getting assets and services out to people, then **reclaim** means taking them back when they're not in use. It's a key part of ITAM. In fact, it's a key part of your return on investment (ROI), and we're going to cover exactly how to reclaim assets within your ITAM practice.

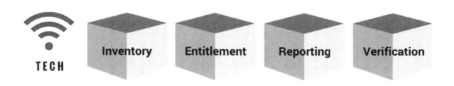

Finally, we have the technology layer. You have **inventory**—everything you have, everything you're using. **Entitlement** is a record of everything purchased and what we're entitled to use and squeeze best value out of.

Reporting is important because it allows us to show progress. We can go back to the authority and say, "Yes, we're doing what you've asked of us." Reporting helps us demonstrate that we're executing our business plan, helps us keep the team motivated on dark days, and helps us keep the stakeholders evangelised and excited about our progress.

The final box is **verification**. If you're not verifying the quality of your ITAM practice, then other people in IT will trip you up, or an external audit will trip you up. You need to verify the data before they do. Verification de-risks your ITAM practice and also increases its reputation and standing.

ITAM IMPLEMENTATION PHASES

I've split the 12 areas into three phases: preparation, project, and business as usual. These phases are worth considering, especially if your practice is starting from scratch.

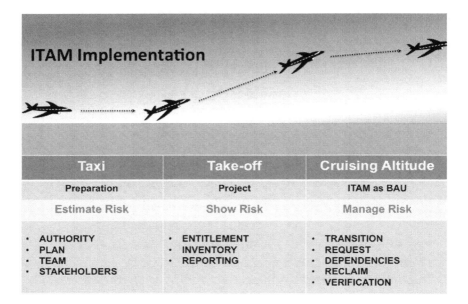

As shown in the diagram above, ITAM implementation is similar to a plane taking off:
- TAXI – Prepare and gather support
- TAKEOFF – Stand up the ITAM department with a project
- CRUISING ALTITUDE – Run the ITAM department in a business-as-usual manner

You need to get all the people things sorted before you get out of taxi mode. You need the authority, the business plan agreed, to recruit a team, and to work with stakeholders. Only then can you go onto project mode: look at entitlement first, then at collecting inventory, and finally at reporting. Reporting will allow you to go into cruising altitude mode, which is when you can really embrace the processes and verification.

In Chapter 2 – Plan, we'll go into "takeoff" in more detail and explore in which sequence all areas of the 12 Box Model should be considered.

In short, the model is a way of breaking up a complex subject into manageable chunks. Throughout the following chapters, we will look at each of the boxes in detail, and consider real examples and case studies regarding how you can use it in your own ITAM practice.

Section 1: Taxi

Chapter 1 - Authority
Chapter 2 - Plan
Chapter 3 - Team
Chapter 4 - Stakeholders

Chapter 1

Authority

SECURING AND MAINTAINING SENIOR MANAGEMENT SUPPORT

For an ITAM department to be successful, a focus on securing and maintaining senior management support is crucial. Without the consent of senior executives, ITAM teams end up banging their heads against the wall and making little progress. Many projects and fires that need extinguishing are competing for senior executives' time. It is critical that we keep the risk and value of our ITAM program front of mind.

ITAM Review research suggests that the biggest blockages to senior management support for ITAM are the following: 1) Higher-priority projects take precedent and 2) The organisation doesn't understand the value of ITAM.[1]

ITAM is often considered a "nice-to-have", an improvement in business processes and housekeeping, rather than a mission-critical practice that will keep the business running. For this reason, frequent communication and justification is required to ensure it gets the attention it deserves.

---

[1] http://www.itassetmanagement.net/2016/02/23/higher-priorities/.

> *"Our senior management feels there are many higher priorities, but I feel they aren't properly prioritising it because they don't understand* ITAM*."* ~ Gemma

We need to build a relevant and compelling yet realistic business plan that wins senior management approval and maintains it over time. You don't want to make a big splash and then fade away whilst IT wanders away to the next shiny project. Keep up the momentum and keep the focus on ITAM.

WHY DO WE NEED AUTHORITY?

Without authority, you're just shuffling papers and not actually making progress.

If you're guiding an organisation from A to B on its ITAM journey, from low maturity to world-class IT asset management, you're effectively delivering organisational change. By doing it properly, you will affect the habits of everyone who touches IT—from the end users using it for their daily work, to those supporting it, to those buying it, to those providing the infrastructure it sits on. ITAM influences everything. By building an ITAM program, you will change how people perceive and value IT assets.

For these reasons, you need leadership behind you to support your program of organisational change. Authority comes in different forms; it might be a senior leader, decision maker, or some other form of authority to help you implement the change in habits required to improve. The authority will help you secure budget and approval for programs, unblock stalwarts stuck in old habits, thump the table when necessary to get things moving, and generally fly the flag for ITAM.

> **Introducing Stella, Eurolager Plc.**
>
> I'd like to introduce you to a friend of mine called Stella. She's an asset manager for Eurolager Plc. (An entirely fictitious company). Stella has recently been recruited to lead an ITAM practice at the company (she was headhunted for her expertise and track record).
>
> The ITAM department she's just been hired into isn't in great standing. It's seen as administrative and bureaucratic. It hasn't got a great reputation within the rest of the IT department.
>
> The company has been feeling the pain of recent software audit activity, which is partly why Stella has been brought on board. An external software audit happens when a software publisher suspects you are using more than you have paid for. Eurolager Plc. is also acquisitive, with acquisition and merger activity in the last year and more acquisitions on the horizon.
>
> Historically, Eurolager's ITAM department is considered policy heavy. In other words, the current ITAM team are too willing to lay down the law to say what is right and wrong, and to lay down policy. Yet, they're unable to defend audits; they lack the basic information to defend the highest priorities. They're seen as a bit of a blocker within the IT department.
>
> Also, by means of background, the IT department at Eurolager is looking to explore mobility in a big way. They've got some significant projects on the go in terms of remote working and digitising the whole business. These are considered headline strategic projects to support the business. Stella's IT department is exploring a cloud-based service desk to provide both support and workflow to support these key projects.
>
> Throughout this book, we're going to accompany Stella in her journey as she attempts to increase the maturity of and make a difference in her ITAM department.
>
> In terms of building authority, Stella plans to create a business case based on her current situation and sell that plan to senior management and the CIO. She also plans to maintain momentum so that she consistently achieves senior management awareness going forward.

ITAM BUSINESS BENEFITS: COMPLIANCE, EFFICIENCY, AND AGILITY

There are many benefits to exploring ITAM, and the most compelling will vary for everyone reading this book. But broadly speaking, the benefits of ITAM can be grouped into three buckets: compliance, efficiency, and agility.

I urge you to consider the business benefits as an à la carte menu. This is not a template; pick and choose which suits your company and situation. The more bespoke your plan, the more impact it will have.

- **Compliance** – Are you addressing the regulatory, legal, and contractual requirements of your assets?
- **Efficiency** – Are you getting best value or ROI for IT assets?
- **Agility** – Do you have compliance and efficiency nailed? If so, you'll possess some fantastic data to help the IT department deliver faster, respond to customer requirements quicker, and make smarter decisions.

Let's explore each of these in turn:

Compliance

The most compelling business driver in terms of compliance is mitigating the risk of software compliance audits. Put simply, if you do ITAM properly, you'll have fewer audits, and those you do endure will be simpler and easier to swat away or manage. Fewer audits means fewer surprises, less unbudgeted spend, and fewer crucial team members being dragged away from key IT projects to search for data to defend audits. Beyond software compliance, your company might also have to adhere to environmental regulations about hardware, contractual compliance or information security.

Efficiency

From an efficiency point of view, there are three key drivers to consider. Think about which apply for your company and select them for your business plan accordingly. Firstly, the big one is the opportunity for cost avoidance. Cost avoidance might be defined as the difference between what you actually paid versus what you would have paid if you'd kept your old habits and ways of buying. This is a massive efficiency area for ITAM, both in terms of hardware and software.

Secondly, there's the opportunity to re-architect. When you really get to know what you have, and how it's configured, you have an opportunity to be more efficient by re-architecting your IT systems to be more efficient—the same value and service delivery using smarter infrastructure and less spend.

Finally, there's the opportunity to renegotiate. You can renegotiate contracts based on real, solid information.

Agility

With better information, we can make smarter and faster decisions. We can be more agile. As mentioned, when we effectively address compliance and efficiency, we have access to fantastic, accurate information about the IT assets we own. We could support new security initiatives, help the service desk be more responsive, help improve the user experience, and support customers better. We could underpin things like migrations and new operating systems. We could support projects and build new data centres and applications. We could provide great transparency in terms of finance. The possibilities are endless.

These days, many vertical industries are entirely underpinned by IT. So, competitive differentiation and business edge might be demonstrated by our ability to deploy new IT systems and services in response to customer requirements. Those who can respond quicker and innovate win. Agility matters.

HOW DO I BUILD THE BUSINESS CASE WITH NO DATA?

New IT asset managers often deal with a paradox. They are typically recruited to build an ITAM practice when asset data is poor, yet in order to build a robust business case, they need good data! If they do have data, nobody trusts it. They have nothing credible in which to build their business case.

Don't panic. There is data to be found in your company to support your business case even when, at face value, it appears you have no data whatsoever.

Three key components of a compelling business case

There are three elements to consider for your ITAM business case: the data you have access to right now, the cost and forecasted improvements, and the business-as-usual metrics.

1. **Right now** – You can harness existing procurement data to support your business case. How much money are you spending annually on IT assets and maintenance? With whom do you spend the most? With which suppliers can you make an impact? It is much more compelling and realistic to base your business plan on a foundation of actual run-rate spend.
2. **Forecasted improvements** – What do you forecast will happen if you do ITAM properly?
3. **Business-as-usual metrics** – What will things look like when things are established? Two years from first standing up your ITAM department successfully, what metrics will be measured to prove you've got an outstanding ITAM department? At the start of your business-plan journey, think about the metrics you're going to measure on an ongoing basis. It's powerful for focus and getting your stakeholders on board.

What is Stella's business plan?

Stella has been at Eurolager for a little while now. She's done her preliminary investigations and is starting to build her business case. What is she up to?

First of all, she's got her vision together. She built a key objective based on the current situation; this is her objective, or her vision for the ITAM department:

"The ITAM department will provide inventory of all assets with 95% accuracy, and an audit-ready status will be maintained for the top ten strategic software publishers with 97% accuracy."

Now, anyone with any experience in the ITAM market will realise that this goal is not easy, no mean feat.

Stella has a very narrow focus, a very tight scope. This is a business plan that is hitting audit defence squarely in the face. Obviously, Stella is missing out on many other benefits of ITAM with this scope. But because she's just starting out, she wants to be really focused. This is a great way of kicking off audit defence, in an audit-centric business plan. As mentioned, you need to pick a focus that suits your current situation. Is compliance more important? Or efficiency? Or agility?

If you're more concerned about cost reduction, or removing waste, or cutting the cost of software and hardware than you are about audits, then the focus in your vision should be slightly different. Just pick something that everyone can measure and see. Every stakeholder that Stella now engages with will be able to see exactly what she's aiming for and whether she is on target on not.

Stella is positioning her business plan to remove the overhead required for audits. The plan is about the cost of audits, the true-up cost, and the audit settlements. It's about the overhead in addressing audits.

Stella's plan will mitigate the risk of audit and potentially reduce the number of audit requests. As a result, Stella will have fewer audits and sharper data in which to respond to future requests.

As part of her business plan, she's also listening to the company's activity. She's looking at providing mergers-and-acquisitions due diligence around IT assets. For example, if her company was about to buy somebody else, she could quickly assess what impact that might have in terms of the existing agreements and assets. In this way, she can support the company and the growth strategy.

Stella is also going to empower the service desk, which is a new initiative for Eurolager Plc. It's a fantastic opportunity to work with the service management team. We will dig into opportunities for ITAM and IT service management (ITSM) integration in Chapter 8 – Transition.

Finally, Stella has built her business plan based on historical spend and audit fines. This is what I mean about looking at data that you have at your fingertips right now, not data that is based on forecasts. Data at your fingertips will include annual spend and historical spend on audit fines and true-ups.

You can estimate how much time you spend addressing audits, and how many working days are tied up in audits. This a real, live metric on money that's actually been spent and that you're hoping to reduce rather than a speculative forecast about what you might achieve.

"TALKING ITAM WITH THE CIO": TONY CRAWLEY, SYNYEGA

Tony picked up the award for ITAM Professional of the Year in the ITAM Review 2015 Excellence Awards.

He has a wealth of experience at the rock face, and in terms of software waste, he's stripped hundreds of millions out of organisations. What does he recommend in terms of building a business plan?

View the YouTube video of Tony here: https://www.itassetmanagement.net/2016/01/26/agile-cio/

Key points:
- Take a phased approach and focus on incremental successes.
- Don't disappear into the basement for two years to do ITAM. Report on progress within the first quarter.
- Regularly communicate your successes focusing on brevity, financial impact, and relevance to the business.

MAKE YOUR BUSINESS PLAN RELEVANT

Stella's business case is focused on audit defence; her company has already been audited several times and has a history of fines.

In contrast, if your company has never received a fine, then a plan based on audits may not solicit senior management's support. There's no point banging on about audit penalties and the threat of audit if your company has never been audited.

The threat of audit may be real, but you need to base your business plan on something relevant to why you've been asked to do ITAM in the first place.

> Stella has begun to distribute and circulate her business plan. She's looking to gain CIO approval for her ITAM practice and plans.
>
> She's built a one-page financial impact summary. It's nice and brief, concise, containing only the financial impact. But she's included a detailed appendix, in case anyone wants to dig in and look at the detail that supports her arguments.
>
> The plan is focused on risk removal and financial improvements, because these things are key to the company at the moment. Stella has also got the endorsement of IT senior management stakeholders. She's got their blessing and ongoing support, learned what's important to them, and built that into the business plan.
>
> It's an incremental plan. Her focus is narrow; she is not attempting to boil the ocean. Her plan is to deliver on her narrow scope and then build upon that momentum using early successes. A key part of Stella's plan is regular communication. She will update people on her progress.

STAKEHOLDERS

Your ITAM practice will have many different stakeholders, and each will have a different impact on your plan. How do you go about identifying and working with these stakeholders?

A stakeholder is anyone involved in the life cycle of IT, from the end user logging in and using services to people building virtual machines and data centres, and everyone in between. Obviously some stakeholders are more important to your ITAM program than others, but ultimately, everyone has an effect.

Typically, the people you want engage with are those in procurement, finance, security, networks or operations, and service management. Depending on the size of your organisation, some of your stakeholders might be business-unit leads (for example, if you're an international company with many subsidiaries). Some of those are likely to have local IT autonomy—they might be your stakeholders. Stakeholders might also include outsource partners or suppliers.

You might have product specialists who are stakeholders; for example, a computer-aided design (CAD) department that looks after their own high-end graphics and CAD software. It depends on what your goals are and what your company is like.

Stakeholders are key to the success of your ITAM business plan.

The power of working collaboratively with stakeholders

There's a trade to be had with all of your stakeholders. What ITAM brings to the table is fantastic asset intelligence. If your data is good and accurate, you can provide this fantastic intelligence about what's happening in the IT environment to support all of the different stakeholders. The ultimate aim is meet your ITAM goals. You want ITAM to be baked into day-to-day operations. You don't want to be constantly firefighting, constantly cleaning up the mess. You want the ITAM governance and processes built into everybody's operations. Basically, you want everyone else to do the work for you. That's the ultimate.

For example, if you're looking at managing the service

desk, and there are engineers installing software and doing upgrades for people, work collaboratively with them. Make sure that these engineers are following ITAM processes so that you're not clearing up their mess. In return, provide them asset intelligence on what they're supporting, so they can be more proactive and do all sorts of progressive service management with this kind of data. For me, it's not about governance and beating people over the head with the governance stick. It's about collaboration, and there's a fantastic opportunity for ITAM to reduce the need for heavy lifting and to be proactive.

Influencing stakeholders

You've got to think about what motivates the people you're working with. You don't need to be Machiavellian in terms of your business plan and deciphering people's motivations, but it's always worth giving it some thought. Knowing what motivates people can be very powerful in terms of delivering your business plan, making things work, and influencing people.

WIIFM, WIIFD, WIIFC

These acronyms (a bit of a mouthful) stand for "What's in it for me?" What's in it for my department?" and "What's in it for the company?"

For example, if I'm working with a contact in finance, what's in it for him personally? What's his motivation? What's he looking to do? What's important to him? What's his pain? What's he currently working on? What are his projects? What does he like doing?

Then, what's in it for that whole department? What is Finance currently up to? What are their key projects?

Finally, what's in it for the company? What's the direction of the company? What's important for the company?

If you approach all of your stakeholders with these questions in mind, you'll place yourself in an incredibly powerful position in terms of getting things done and influencing people. This is not about manipulation—if you help people with their goals, they're going to help you with yours.

Some of you might think I'm teaching you to suck eggs here, in terms of working with your colleagues. But I see far too many people hiding behind their inboxes, firing off emails, making demands about what they need for ITAM without actually collaborating with people. To make progress and lasting change to people's habits in the organisation, make ITAM about collaboration, not just telling people what they need to do. .

Social lubricant
I'm all for digital. We're in a modern era where a lot of your daily contact with your colleagues is going to be digital, through conference calls and web meetings. Some of you are probably in international offices with very little in-person contact. However, I still think nothing beats meeting some of these key stakeholders in person if possible.

Beer, coffee, doughnuts, pizza—these are social lubricants. Whatever you might need to get to know people in an informal setting. Find out about what makes them tick, how you can help them, and then they're very likely to help you back. When I look at world-class IT asset managers, that's what they're doing. They're embedding themselves in the industry and in their sector. They're not pinging off emails and telling people off for not obeying the rules. They're making lasting change by helping people change their habits around IT assets.

In terms of influencing stakeholders, consider the "carrot and stick" concept. Historically, ITAM has been very much a stick, a security function, pulling people up on policy. But it's much more enjoyable and progressive if you can use the carrot as well. What's in it for that person to help you? How can you help them? I don't want to belabour the point. I'm sure you see that there's a great opportunity for collaboration.

> Let's return to Stella's journey. What is she doing to build her reputation, work with stakeholders, and build traction with her business plan? Well, she has recognised that there are three key departments in her environment to help her execute her business plan.
>
> Firstly, her company is implementing a new service desk, so she has a great opportunity to bake some ITAM governance into it. Secondly, she will be working closely with operations because, historically at her company, operations have borne the brunt of the audit-defence work. They're all-too-keen to help her succeed because it will free up their workload. Thirdly, she has identified that procurement are key because they are behind a lot of the merger and acquisition activity for Eurolager in terms of investigative work. She sees procurement as strategic in helping to execute her business plan.

Maintaining momentum

How do you maintain momentum with your ITAM practice and ensure your authority continues to support and pay attention to your progress? The first way is the obvious one. It's not unique to ITAM. It's not unique to any discipline within IT, or to business in general.

Just communicate.

Life happens, things happen, other things pop up on the radar. You've just got to keep communicating: "We're still here. We're still progressing. We're still doing all the stuff we promised to do." It's just basic communication.

The second way to maintain momentum is to build a reputation for ITAM intelligence. Leading ITAM professionals build a reputation for excellence around ITAM in their company. That means providing great data, analysis, and intelligence for everybody in IT. Build momentum by being seen as a lynchpin, a great resource, and a great part of the team.

Finally, expand the focus. This might sound obvious, but I'm surprised by how many people get stuck on this. As discussed, an ITAM business plan is based on compliance, efficiency, and agility. Depending on your priorities, you're going to lean on one of those priorities. But if you want to continue your momentum, expand the focus.

I'll give you an example. I've come across quite a lot of organisations that go on a cost reduction drive with ITAM. They strip away software and hardware waste. Doing this for two or three years, they'll deliver massive returns. Then naturally, as they get leaner and more efficient, the returns dwindle. They've succeeded. The momentum in their ITAM functions starts to diminish, and they get stuck. To avoid this, you need to address compliance and then expand into agility: "How can I use this asset data to support other initiatives throughout IT?"

Stella has started to communicate. In order to engage regularly with stakeholders and maintain momentum, she hosts a monthly ITAM board meeting, getting all her stakeholders in one room. This is a fantastic mechanism. By securing a specific date in advance, you can ensure you'll have people's time and attention. It's a great way of reviewing progress, deciding actions, and moving on.

The meeting is about reporting on progress and then taking action. It's not about pinging emails around with Excel spreadsheets that nobody reads, or going into meetings and nodding politely. Make sure that your communications, as Tony mentioned in our case study earlier, are financially focused. It's easy to slip back into ITAM mode and start talking ITAM language when it's not appropriate.

Stella and the CIO have agreed that the CIO will not attend all of these meetings. He doesn't need to, but he will certainly review the minutes that the meetings generate and ensure the board have a degree of authority and progress is communicated to everyone who needs to see it.

Stella has built specific communication procedures for the service desk, operations, procurement, and the broader IT team. She realises that if she's going to execute her plan she needs the ongoing attention of these departments, so she communicates updates that are relevant to them.

Stella is also keeping a journal of her impact, progress, and savings. It is a great resource to lean upon when reinforcing her message.

If you've nailed software, expand your focus to hardware, to mobile, to data centres. Again, I know it sounds obvious, but you'd be amazed by how many organisations get stuck in a rut when it comes to their scope. They're making good progress, so they don't realise why they're losing momentum.

Summary

When it comes to building a relevant business plan, don't work with a boilerplate template. Think about your company—what's going on, what's relevant, what's hot; what are the key words, the key initiatives? Make sure that your ITAM aligns with your company.

Look at the senior management influence. Look at the language that they're using, and the brevity of it. Then make sure your messages are on point.

And always think about momentum. How are you going to keep up the momentum of your ITAM practice? How are you going to secure authority and reputation?

In the next chapter we will explore how to harness senior management support to deliver a compelling business plan.

Chapter 2

Plan

BUILDING A THREE-YEAR ROADMAP

Once you've gained senior management support, you need to identify and prioritise risks, build a three-year roadmap for your ITAM department, and communicate your success.

ITAM FIRST PRINCIPLES

Before digging into the specifics of building a three-year business plan, it's worth taking a step back and looking at some basics of IT asset management. When goal setting, building plans, and thinking about long-term goals, consider, where could I take this? ITAM is much more than compliance or admin, so let's explore the opportunities.

MONEY MANAGEMENT

In layman's terms, "asset management" usually refers to managing money.

If you own a big pile of cash, you might ask an asset management company to manage it for you. The company will invest your assets in a portfolio of financial products to manage on your behalf. Which products they invest in will depend on your outlook on life and attitude to risk.

A person close to retirement might opt for safe and steady government bonds with a low but predictable return whereas a more adventurous investor might opt to invest in start-up businesses that are higher risk but have the potential for more lucrative yield.

The underlying principle: an asset manager puts your assets to good use whilst respecting your attitude to risk.

AIRCRAFT CARRIERS, FORKLIFT TRUCKS, AND CHAINSAWS

Beyond the world of finance, asset management principles can be applied to everything a business owns, from tangible things like aircraft carriers, forklift trucks, and chainsaws to intangible things like ownership of intellectual property, brand names, and goodwill.

These assets generate a return for the business. The return might be quantified in terms of productivity, performance, profitability, or the best possible service to customers.

Business assets also carry varying degrees of risk. These risks can be classified as:

- Financial (the asset may generate costs)
- Contractual (the asset might be useable only within certain contractual terms)
- Legal or compliance (the asset might be constrained by certain laws or industry regulation)

ASSET LIFE CYCLES

Asset management principles are applied to the whole life of the asset, from the time it begins to have an impact on the business to the time it is removed.

- Cradle: The life of an asset might begin when a business creates the idea of it, includes it as part of a strategic plan, or intends to purchase it, i.e., by raising a requisition or purchase order.
- Grave: The life of an asset might end when the asset is destroyed, when it expires, or when its ownership is passed legally to another organisation.

In simple terms, an asset's life cycle can be broken into four general elements:

- We create or acquire assets
- We utilise them
- We maintain them
- We renew or we dispose of them

It is important to consider the whole life cycle when managing assets, since each element affects the overall risk and value of that asset. The following fictional example is how a company might manage non-IT assets.

LOGCHOP4U INC.

Logchop4u is a tree surgery and forest maintenance company. Chainsaws are a key strategic asset for the company; they are critical to the profitable delivery of contracts and also carry significant risk, from the mundane to the life altering.

The company must strike a balance between maximising the output of the chainsaws and minimising risk.

Logchop4u also faces certain legal constraints when using its chainsaws, such as use of safety boots, gloves, and masks worn by certified licensed operatives.

It employs a chainsaw maintenance firm to service the engines and maintain cutting chains. Sharp chains on chainsaws free from sawdust perform better. The company also maintains central records of all key equipment and staff.

The company has learned that proactive maintenance of their chainsaws allows them to meet their legal requirements and also improves the profitability of contracts. Well-serviced chainsaws operated by highly trained operatives equates to less down time due to broken equipment, prolonged equipment life, and lower insurance premiums.

When a chainsaw comes to the end of its useful life with the business, Logchop4u sells it to another firm and receives a certificate to prove that it is no longer an asset within the business. The company is thereby released from its legal responsibilities relating to that asset.

Logchop4u is able to minimise risks whilst maximising returns from their assets, which equates to better customer service and more profitable contracts. These are asset management principles at work. You might also call this common sense, or simply good business practice.

EMOTIONALLY COLD

It is worth noting that it requires a certain level of emotional detachment to manage something purely as an asset.

A house and a car are two of the most expensive items an individual is likely to purchase. Yet very few people "sweat" these expensive purchases as much as they could. Very few people rent their cars out when not using them, or sublet their houses for alternative use during the day. These items are, traditionally at least, seen as personal; they have emotional value.

You might think this principle applies only to personal purchases and not to those in the business arena—but many strategic business decisions around assets are inextricably linked with company politics, history, and emotional attachment.

For example, Ford Motor Company announced their decision to cease production in Australia. If the Ford Australia territory is viewed as an asset in the broader Ford portfolio, then this decision is a no-brainer (lost money for the last five years, cheaper to import cars into Australia than manufacture locally). However, this was a wretched decision for Ford to make. They'd operated in the country for 85 years and provided 1,200 jobs.

Unless you're a quartermaster in a Buddhist temple, you'll have some degree of emotional attachment to your assets. Emotional attachment is apparent within the largest decisions (leaving a territory) and the smallest (buying a loaf of bread). Those of us working in the asset management field need to be sensitive to these connections.

PRIORITIES AND CONVENIENCE

Asset management can be defined as:

> *"Any system that monitors and maintains things of value to an entity or group."* ~ Wikipedia

The asset management "system" requires horsepower to run in terms of human input, processes, and checks and balances. This administrative housekeeping takes effort—therefore, every business has to prioritise which assets to manage in order to make best use of their time and resources. The law of diminishing returns applies: does the cost of managing the asset outweigh its value?

It is said that behaviour in the business world is ultimately governed by money and legislation. The same can be said of asset management—convenience and financial return are at play. Assets get managed when legislation dictates, when it's convenient, or when financial rewards make it obvious to do so.

The point of exploring asset management first principles is to show that ITAM is a broad topic and not just about compliance. ITAM is an area of huge opportunity to organisations that are willing to invest in its implementation. From an industry maturity perspective, we are in the foothills. It's worth exploring the true potential of ITAM as we plan our future.

MAJOR MILESTONES

There are five major areas to consider when building a three-year ITAM roadmap:

- Visibility – Address, as Donald Rumsfeld[2] might say, "known unknowns" and "unknown unknowns" in your environment
- Compliance – Maintain audit-ready status for your most strategic publishers

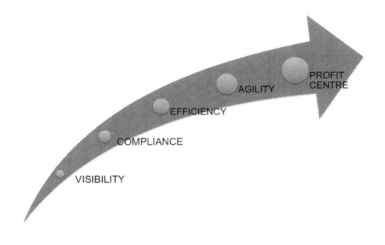

---

2   https://en.wikipedia.org/wiki/There_are_known_knowns

- Efficiency – Drive optimal value and lowest cost from all IT spend
- Agility – Provide the IT department with trustworthy asset data to enable faster decision-making and project support
- Profit centre – Allow ITAM to pay for itself, be self-sustaining, and create its own budget

> **How will Stella achieve her goals and deliver a three-year business plan?**
>
> As we learned in the previous chapter, a key part of Stella's plan is based on addressing compliance. Stella's peers or competitors at other companies may wish to focus on other things, but for Stella, compliance is the main objective.
>
> Stella is conscious that her CIO is juggling many other priorities demanding attention. So her plan needs to be relevant, compelling, and realistic. Her plan needs to make her senior management team sit up and pay attention, yet it needs to be based on facts or credible research.
>
> Money talks. Stella needs to strip away non-essential information and focus on financial and commercial impact.

A LACK OF VISION

Professionals embarking on their ITAM careers often ask, "What metrics should I track?" This question suggests a lack of long-term vision. Once a long-term vision is identified, it becomes obvious which metrics to track: metrics that demonstrate you're moving towards your goal.

ITAM *implementation phases*

When building your three-year ITAM roadmap it's worth considering the different phases to ITAM implementation. This is especially true for a company starting from scratch. The three phases are preparation, project and business as usual.

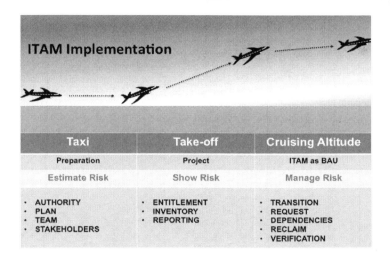

As shown in the diagram above, ITAM implementation is similar to a plane taking off:

- TAXI – Prepare and gather support
- TAKEOFF – Stand up the ITAM department with a project
- CRUISING ALTITUDE – Run the ITAM department in a business-as-usual manner

In the market, we often see "planes" prepare and take off only to descend gently to earth. Momentum is lost. It's vital that your three-year plan incorporate communications and reporting to ensure momentum is maintained and your ITAM practice remains airborne and successful.

WHAT SHOULD YOU FOCUS ON FIRST?

The diagram below suggests a potential sequence in which to address all 12 areas. This is not prescriptive, and should be adapted depending on your priorities, but it's a good sequence for an ITAM practice being built from scratch.

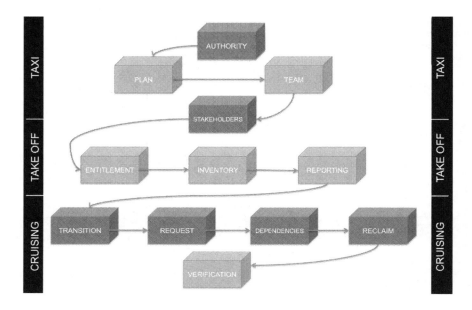

WHO TO TRACK?

In Chapter Six – Inventory, we'll cover the Pareto principle and the benefit of focusing efforts on a select few publishers in order to maximise use of resources.

For now, to select strategic software publishers for the ITAM department to manage, ask yourself the following questions:

1. **Risk** – Which vendors represent the highest risk? Which ones have audited us most frequently? Which ones are likely to audit us based on market activity?
2. **Spend** – Who are our highest-spend publishers per annum?
3. **Volume** – Which vendor has the most software in our environment?

4. **Strategy** – Which suppliers are strategic to our business in the longer term?
5. **Transition** – Are we moving towards or away from certain suppliers in the next few years? Would it be prudent to get our house in order for those suppliers as part of our plan?
6. **Gut feel** – What is my personal opinion and that of my IT department colleagues as to who should be our key supplier to track?

After thinking about these questions, you should have a good idea of your top ten to twenty vendors. Part of your three-year plan can include expanding your list of top vendors based on your progress and success and, if necessary, recruiting more resources to expand your team.

> Stella is letting the numbers do the talking. In order to reinforce the scale of the problem, she simply calculated total spend for her top ten vendors and shared this information with the senior management team. This helped illustrate the sheer size of the issue. Total spend really made Stella's stakeholders sit up and think about how much money they were spending on software and why it was important to manage it. This is why we need to do ITAM. She has also made her business plan iterative in that she plans to first address Microsoft as a publisher, deliver results within a few months, deliver an impact, and then move on to other publishers. Stella also recognises that no business plan survives first impact with the customer; an iterative plan is flexible, able to change based on feedback and commercial priorities. Things happen, life happens, and your plan needs to flex and adapt over time.

COMMUNICATIONS

Finally, as part of your three-year roadmap, consider how you will maintain the momentum and communicate your progress.

You have to keep the drum beating because no matter how compelling and exciting your plan is and how much money you're going to save the company, other things will happen. New projects come up, attention wanes. Keep the focus on your plan by keeping the communications going.

Think about your main audience, typically end-user customers, internal IT teams/stakeholders, and your senior management team. How will you communicate your activity?

Communications channels to consider:

- Lunches/informal training
- Internal email
- Newsletters
- Posters
- Forums, social media, noticeboards
- App Store/online portals (communicate the cost of software to reinforce why you're managing it, and also communicate your willingness to reclaim the asset if it's not in use)
- Life cycle opportunities

These are just some of the channels we've seen ITAM Review readers use over the years. You are only limited by your imagination. Speak to your internal communications team and see what works for them.

With continual updates and communications you can keep the plane in the air and deliver your plan.

Chapter 3

Team

BUILDING A WORLD-CLASS ITAM TEAM

In the three-phased approach to ITAM take-off, "team" sits squarely within the taxi (preparation) phase. We've developed the authority and we've built the implementation plan. Now, we get the team together to help us execute.

In this chapter we'll cover how to determine what's required in your team, how to assess their skills, and how to develop a team that will allow your plan to succeed.

ITAM THE SUPERHERO

IT asset managers require many skills and qualities. This manager must be:

- **a negotiator** securing the best deals;
- **an Excel guru and data analyst** recognising truth in masses of data;
- **an investigator** working with sparse information and threads of evidence;
- **a translator** turning financial, procurement, and technical information and business demands into a common language;
- **a mediator** understanding the needs and wants of all parties and seeing the bigger picture;

- **a fire-fighter** dealing with emergencies and last-minute requests;
- **a gatekeeper** defending policy and keeping costs low;
- **a project manager** delivering new systems or approaches;
- **an ITAM tool wizard** driving best value out of ITAM technology;
- **a librarian** maintaining accurate records in an orderly fashion;
- **a communicator** talking to everyone in the business, from the everyday IT user to the senior management team;
- **a licensing guru and contract specialist understanding licensing terms and product use rights**; and
- **an agent of change** making things happen.

As you can see, it is quite a diverse and varied list of requirements. I hope you like a challenge! These are also skills that don't necessarily lend themselves to one personality type. For example, a person ideally suited to being a librarian might not be an excellent global communicator. The role of IT asset manager can be diverse, challenging, and rewarding, with plenty of opportunity for specialisation and future career opportunity.

ITAM is varied and far-reaching. You could line up two excellent candidates for your team and they might have two entirely different skill sets. Job titles and roles are just discussion points and shouldn't be taken too literally. Ultimately, you just need to build a team that meets the plan.

ITAM: A UNIQUE AND VERSATILE ROLE

To help you understand the ITAM role and ITAM team structures, let's explore some existing ITAM job titles and how they might hang together to build a team.

To assess the different job titles in the market today, I downloaded a subset of ITAM Review–reader job titles. I selected 2,500 readers at random and found 1,726 different job titles. Clearly there is a great deal of diversity within ITAM roles. The following are the top ten most popular ITAM job titles:

1. IT asset manager
2. Software asset manager
3. Consultant
4. Project manager
5. SAM analyst
6. IT manager
7. IT asset analyst
8. IT asset administrator
9. SAM consultant
10. Software licence manager

So if you're building a team and looking to hire staff from the pool of worldwide ITAM talent, you might consider these roles.

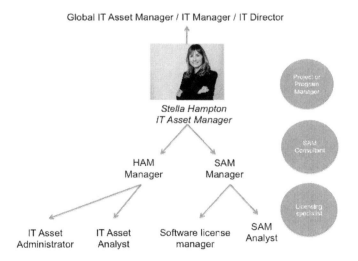

> ### Traditional ITAM hierarchies
>
> With these top ten job titles in mind, what can Stella do to build her team?
>
> Stella is an IT asset manager, so if she takes the traditional route, then she might be in charge of the hardware asset manager and the software asset manager. Then within hardware, there might be an asset administrator, to keep records up to date and handle administration, and an analyst, to help with forecasting, hardware requirements, etc. There might also be a software licence manager and SAM analyst within the SAM team, all of who might report to the IT asset manager.
>
> Finally, Stella could sprinkle some tactical requirements into her team. For example, if Stella rolls out a new SAM tool, or a new process, she might hire a project or program manager. Or if she wants an external expert to say, "How can we improve things?" she could hire a SAM consultant. She might also have some specialist licensing expertise brought in to help with her top vendors.
>
> Straightaway, she's built a hierarchy. Stella herself might be part of this extended team as well. She might report to a global IT asset manager. This manager would typically have global remit, and IT asset managers in various different territories (the Americas, Europe, and Asia-Pacific, for example) would report to them. This is an example of a fairly traditional ITAM team; we'll learn later how Stella specifically goes about building her team..

HOW BIG SHOULD THE ITAM TEAM BE?

You've just seen how we could structure a team of ten for Stella, but how big should your ITAM team be? My simple answer is this: it should be big enough to deliver what you've promised in the business plan. But it's often useful to look at other organisations.

> ### ITAM and the career path to CIO
>
> In terms of career path, Stella is in a good position for future growth. If she's able to execute her business plan, she'll be a change agent—someone who can build a powerful program of change to help the company progress.
>
> Stella is a communicator: she works with every area of IT, and she communicates in terms of business benefit and commercial impact. This also puts Stella on a path towards the CIO position, should this be something she wants to consider. She'll obviously need experience and reputation, but ITAM is a great opportunity to deliver that.

The image below comes from the 2014 ITAM Review salary survey.³

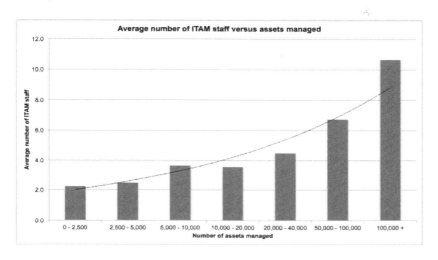

We asked two questions of ITAM Review readers:

1. How big is your team?
2. How many assets do you manage?

The graph above shows the approximate relationship between team size and number of assets managed. So, for example, organisations with 10,000 to 20,000 assets typically have three to four staff in the ITAM team. This is a couple of years out of date and the ITAM discipline is evolving, so teams are likely to be bigger in the future as more companies realise the full benefits of a mature ITAM practice. But it's a start—a quick guide to how your team compares to the rest of the industry.

---

3   https://www.itassetmanagement.net/2014/06/11/salary/.

WORLDWIDE SKILLS SHORTAGE

Another consideration when building an ITAM team is that, at the time of writing, there is a worldwide skills shortage for ITAM. Building a team might be a challenge. Organisations are struggling to hire and retain experienced ITAM talent. Skills in most demand are software licensing, software asset management, and contracts experience.

Wages are going up considerably, and it's a sought-after role. If you're lucky enough to be a software asset manager, you've likely got a great future ahead, professionally.

> **Building a team to execute the business plan**
>
> So how do we apply this knowledge of teams and job titles to Stella's business plan requirements? By going back to her vision.
>
> As mentioned previously, whilst it's sometimes useful to look at standards and other organisations' teams, ultimately a team should enable the IT asset manager to deliver her vision.
>
> Let's return to authority in the 12 Box Model. Stella's plan is focused on defending against and preventing future software audits, as these are painful and costly exercises for her company. It's all about reducing the audit overhead in terms of managing audits, reducing the risk of audits, reducing the spend every year on audit true-ups and fees, and so on. Also, her company is acquisitive, so she wants to provide good asset intelligence to help the due-diligence process for merger and acquisition activity.
>
> To articulate this requirement to her senior management team, she boiled it down to a single statement:
>
> "The ITAM department will provide inventory of all assets with 95% accuracy, and an audit-ready status will be maintained for the top ten strategic software publishers with 97% accuracy."

The role of service providers and external resources in teams

Service providers and specialists also play a role in the team, and it's common for teams to include elements from external parties.

ITAM managed service providers have become more popular in recent years; especially SAM managed service

> **Key requirements of Stella's team**
>
> Stella's statement can be broken into chunks for the team to manage. Stella needs to ask herself, "What are the key requirements for the team when it comes to delivering my vision?"
>
> At a high level, Stella needs trustworthy inventory from the team. She also needs an effective licence position (ELP) for her top ten vendors (ELP is covered in Chapter 5). But there's no point just delivering accurate inventory and the ELP every month—she wants things to progress, to be better. So Stella needs to include an element of refinement and continual service improvement (CSI) in the team responsibilities (Reporting and CSI are covered in Chapter 7).
>
> Again, I'm not suggesting that Stella's way of building her team is the way you should build your team. But I would recommend a similar process of setting a specific target in regards to the senior management team and aiming the whole team at that target.

providers who provide compliance visibility for a fixed monthly fee. Some organisations choose managed service providers as a short-term fix whilst they build out the capabilities of their team; some choose to work with managed service providers as a long-term solution to allow their team to focus on more strategic ITAM initiatives.

Managed service providers are outcomes focused. In other words, rather than buying a tool (a bunch of consulting and services), you buy an outcome. Managed service providers or consultants also provide coverage for any holes in your team's capabilities.

> **Stella's team plan**
>
> Stella's got three high-level requirements: inventory, an ELP for the top ten, and CSI to ensure ongoing refinement and make things better.
>
> What could Stella do to build out her team to better deliver her plan?
>
> Stella has got Petra, an ITAM analyst, doing the inventory. Petra is solely responsible for trustworthy inventory at 95% accuracy. That is her number one priority. Doing inventory can be a Pandora's box experience—you wish you hadn't opened up the box because it revealed issues with all of the systems and assets.
>
> Petra's immediately hit a bottleneck in terms of constraints and technical resources. She's looking to hire someone to help her deliver the accurate inventory because it's so incredibly valuable for everything else that's happening. That might be a dotted line to a resource in somebody else's team. It might be a contractor, it might be a full-time resource, but that's something for Stella to resolve.

Stella's Team

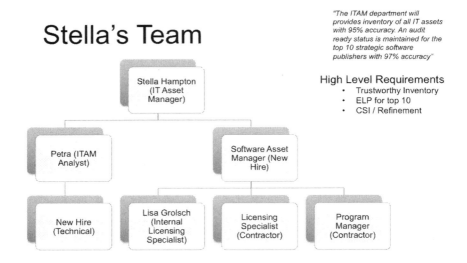

"The ITAM department will provides inventory of all IT assets with 95% accuracy. An audit ready status is maintained for the top 10 strategic software publishers with 97% accuracy"

High Level Requirements
- Trustworthy Inventory
- ELP for top 10
- CSI / Refinement

On the software side of things, Stella's identified that she needs a software asset manager. That's a new hire. Lisa, who's already in her team, has expressed an interest in progressing into that role, which would be a great move for her.

Stella also needs a licensing specialist (either temporary or long term), because although Lisa's very good, she can't cover every single licensing model. As well, Stella sees the need for a program manager due to all the change going on (for example, the new service desk and new SAM practices and processes).

In terms of responsibilities, we've got trustworthy inventory going to Petra and ELP responsibility going to the new software asset manager (hopefully that's Lisa). Then, we've got the overall CSI and refinement, which is going to Stella and her team. This way, everybody is focused on the vision and what the team is looking to achieve. This might be management 101 for many of you, but it's how you might go about hitting your vision and business plan.

TEAM BUILDING 101

The following advice on team building is not restricted to the ITAM field. People in all fields want to see progress; they want to be part of a department or a team that's delivering against goals.

Stella's peers and extended team

We also need to be conscious that Stella isn't just managing her direct reports; she's also managing the expectations and requirements of her peers, stakeholders, and extended team.

Stella can't deliver her business plan for ITAM in isolation with her team. She needs to communicate and work with the rest of the business. As well as working downwards with her team, she needs to work upwards and across with her peers and senior management team. There's James, the service desk manager, Angela, the enterprise architect, and so on.

Stella also needs to work with her regional IT managers in different territories. It takes only one rogue IT manager in some territory to go off doing their own thing to potentially disrupt the whole business plan.

Stella also has to work with stakeholders outside the IT department. In this instance, procurement is reporting upwards to the CFO, and she needs to work with them. Then there are the different business units and department heads, as well as the end-user IT customers.

Ultimately, Stella's influence is going to touch every one of these stakeholders. If you're doing ITAM properly, if you're changing the way people touch IT, you're ultimately changing the habits of everyone on the org chart. Stella is going to need their cooperation and support.

Eurolager Plc.

Stella Hampton
IT Asset Manager

James Peroni,
Head of Service Delivery

Petra Hofbräu
ITAM Analyst

Stuart Becks
CIO

John Amstel
IT Procurement

Lisa Grolsch,
Licensing Specialist

Rallying around a common goal is great for team building. Career and financial progress might be a bit more important, but your team also wants to make a difference, to make a dent in the universe and deliver progress.

Exposure to new experiences
Having spoken to IT asset managers in the field, I've found that industry-leading ITAM professionals are seeking exposure to new technologies, licence models, territories, challenges, and opportunities. They want to keep learning and adding more experience to their portfolios.

Increasing scope
The final thing to consider in terms of team building is growing the scope. For example, Stella's business plan is very much focused on audit defence and reducing the threat and distraction of vendor audits. To develop her team and retain good talent, it would be wise for her to expand the scope of her team once audit-ready status was maintained.

Stella could look at cost reduction and removing waste, or re-architecting systems to drive value and lower the cost of computing. She could look at security, or supporting a move to a new platform or operating system. Expansions in scope will provide new challenges and experiences for her team to sink their teeth into, and will provide more value for the ITAM department.

Summary
In this chapter we covered how to assess the requirements of your team, focus on the business plan, break up the business plan into manageable chunks, and distribute responsibilities across the team. We also covered the current state of ITAM careers and how service providers and

external consultants may be an option to fill in the gaps in your requirements.

We looked at how to manage ITAM responsibilities downwards in a team as well as upwards and across in an organisation. We also touched on how to grow a world-class team in terms of developing potential. Ultimately, I would want to be part of any team that's progressing, making a difference, and giving me the opportunity to grow.

Chapter 4

Stakeholders

INTRODUCTION

In this chapter we explore how to use ITAM data across the business, outside the IT department, to build your ITAM team reputation and influence.

Better standing in terms of reputation and influence leads to more resources, more team members, and more impact. In short, it leads to a powerful ITAM department that is able to make progress and make a difference.

Read this chapter to learn how to get stakeholders involved, how to work with them, what to do when collaboration doesn't go so well, and how to use ITAM data beyond the ITAM department.

WHY BOTHER WORKING WITH STAKEHOLDERS?

Throughout the book I've emphasised the role of stakeholders. I don't believe you can do ITAM without proactively working with them.

As a tangible example, let's look at a very specific process in a specific organisation: reclaiming kit when an employee leaves. Below is the RACI chart for the task "Reclaim and process IT equipment from employees leaving the organisation".

STAKEHOLDERS

Task	Responsible	Accountable	Consulted	Informed
Reclaim and processing of IT equipment from employees leaving the organization	First line support	Head of Service Delivery	ITAM, Security	HR

What is the **task**? When an employee leaves, all the kit they've been issued must be clawed back and processed (cleansed, rebuilt, sent to outsourcer, etc.).

- Who is **responsible** for performing the task? First-line support; in this organisation, the task is initiated by an incident or ticket from HR.
- Who is **accountable** if things go pear shaped? The buck stops with the head of service delivery.
- Who should be **consulted**? Both the ITAM team and security.
- Who should be **informed** when the process is complete? HR, as in this organisation the kit must be returned before HR can close their internal HR leavers' process.

So whilst ITAM is consulted, our IT asset manager is not responsible for the success of this process. A successful reclaim process is critical for things like cost avoidance (stockpiling assets to avoid purchasing), and yet the ITAM team has NO control or responsibility over this process.

Alternatively, of course, you can say, "Forget it, that's their problem not mine." But that won't allow you to make progress. The goal is to make LESS work for yourself and stop the barrage of errors and exceptions coming your way every month.

This is why it's critical to work with stakeholders. The success and progress of your ITAM department is dependent on it. Anyone who can affect the success of your progress, processes, or key metrics should be considered a stakeholder. In this example, the IT asset manager needs to empower and support first-line to do their job properly.

IDENTIFYING STAKEHOLDERS

Anyone who's going to have an impact on the success of your ITAM project, processes, or metrics is a stakeholder. And certain stakeholders in your IT department and business will be key contacts.

There is no cookie-cutter template when it comes to whom to work with. It depends on who is doing what within your organisation. It also depends on personalities; certain individuals will have influence that transcends their job titles.

Typically, you'll be working with the likes of procurement; security; network operations; whoever is responsible for looking after devices in both the desktop and the data centres; service management and the help desk; business-unit leads; and product specialists. If you're in a large company with multiple territories, you might be working with lots of different IT departments—they might be stakeholders. You might have enterprise architects and project managers. You might have outsourcers as stakeholders. Again, consider who is going to influence your metrics; who is going to influence your success?

COMMUNICATING WITH STAKEHOLDERS

Reporting is very important to stakeholders; a whole chapter is dedicated to reporting. It allows you to demonstrate that you're delivering against your business plan, to present the high-level risks and trends to stakeholders, and to keep them interested.

Reporting allows you, via trustworthy data, to continually improve your reputation and remind stakeholders of the value of ITAM. We need to constantly remind ourselves that ITAM is a nice-to-have function. It's incredibly valuable and delivers an incredible return, but it's still a

> **What do stakeholders want to see?**
>
> If we look at Stella's peer group we can see the different requirements for ITAM data from different stakeholders:
> - James, the head of service delivery, has a concern about application version sprawl and device sprawl. His team has only so many hours in the day but is being asked to support an ever-increasing array of applications and devices. His team's being spread too thinly, and this is a real concern. Stella can support him with accurate device and application variety information, detailing device volumes and application version diversity.
> - Angela, the enterprise architect, is looking at future options for SQL Server, in terms of what the technology roadmap for this application should be. Stella can help her by working with the procurement team to provide a vendor roadmap (what versions will be released when) and providing information about existing deployments and when they will no longer receive critical updates and patches from the vendor.
> - Some regional IT managers within Stella's network are also planning further integration with the central IT team. They can work with Stella to understand what products are used where, what the organisation could standardise on, and, in conjunction with procurement, what the associated costs might be.
> - John over in procurement is particularly interested in the usage of devices and applications, customer satisfaction in regards to devices and software, and general consumption because this information aids the contract-negotiation process.

nice-to-have. It's something that people ignore if you stop reminding them of its value. So communicating with stakeholders is very important.

You need to think about your own network—your peer network of stakeholders and internal influencers. What are their key concerns and how can you help? Ultimately it's about listening to what's going on in the environment. It's about building relationships with these people, communicating regularly, and ensuring that communications are relevant and accurate.

As mentioned, the glue that holds stakeholders, reporting, and progress together is some form of ITAM board, steering group, or committee. An event in the calendar whereby stakeholders get together and review progress is a great way to maintain momentum. An agreed action plan ensures stakeholders are motivated and on the same page.

BEING PROACTIVE WITH STAKEHOLDERS

At the time this chapter was written, news broke about Internet connected devices being used in a cyber attack: https://www.itassetmanagement.net/2016/10/26/iot/

In a nutshell, some cameras were Internet connected and had a vulnerability that allowed a third party to take them over for malicious intent. The compromised cameras were coordinated to work together to attack certain companies and networks.

This is the sort of thing that might pop up weekly or daily in terms of news for security professionals. And we as ITAM professionals could potentially have some influence over this.

We could think, "Okay, well, we've got our discovery data. We probably haven't got an inventory agent installed on a camera because that's not appropriate, but we can probably identify where it is and maybe even get some firmware details. Perhaps we could identify some intelligence about what might help security address this issue."

Maybe your security team has got this risk nailed. Maybe there's a strong firewall in your organisation to protect any outbound traffic from cameras etc., but this is an example of something you could approach your security team with; you could say, "Look, do you want me to run a report based on this?" Or, "We've got some good data on this." Think about your stakeholders and what might be valuable to them. Again, it's about being proactive. You've got to consider that your stakeholders' requirements are ongoing; they're evolving and changing over time, not set in stone.

STAKEHOLDERS AND ESCALATION

What happens when things aren't going so well? We've talked about pizza, beer, and building relationships, but what if that relationship isn't there? What if your stakeholders aren't listening? What if people aren't turning up to your meetings or reading your reports?

Let's go right back to basics: the role of an asset manager—any asset manager in any industry—is about risk. Your role is to manage risk first and foremost, so it's your responsibility to make people aware of it. If they're not listening, you need to escalate it. You need to say to the senior management team, "Look, you've asked me to do this role. I'm trying to move in the right direction, but this isn't happening. If this isn't addressed, here are the risks."

When escalating an issue to senior management teams, think about the trustworthiness of your data. There's nothing worse for your reputation (or more embarrassing!) than escalating things that aren't actually accurate. Accurate data is essential.

You also need to be aware of not crossing the line and being seen as a snitch: "Oh god, it's the ITAM police. It's the Department of No." Give people every opportunity to work with you, and escalate things only as a last resort when things aren't progressing.

Finally, in terms of escalation, emphasise the team effort. For example, rather than saying, "I'm trying to achieve this and it's not happening and this person isn't helping me," you could say, "Look, I'm doing this for the team. This exercise is of value to the whole IT team. This is what I'm trying to do for everybody, and this thing is not happening—this is why I'm escalating things."

Escalation is necessary and important, a critical part of your role as asset manager. But remember, senior management want to read concise reports focused on commercial terms and financial impact. Escalations need to follow the same template—brief, with a focus on risk and financial impact.

TRUSTWORTHY ITAM DATA IS GOLD!

Good, trustworthy ITAM data has massive potential across the whole IT department. The mind map below provides a few examples of how good ITAM data can be used outside of ITAM. Each "branch" is a potential project that is leveraging ITAM data but is outside ITAM's core responsibilities.

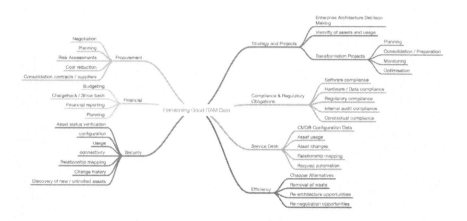

CIO Priorities

Another thing to be cognizant of when working with stakeholders is the CIO's priorities. Let's revisit Stella's approach to seeking senior management approval.

Stella looked at the CIO and company priorities to ensure her business plan is relevant. You're not going to get sign-off on the ITAM budget if you're off on a tangent. Your budget needs to be aligned with what the CIO is up to and what the business is up to.

Stella has identified the top ten projects that her company is working towards and investigating:

1. Modernised infrastructure
2. Modernised/consolidated data centres
3. Cyber security
4. Transition to cloud (on-premise, hybrid, private, public mixtures)
5. Digital transformation and automation of back office

> 6. Service desk automation, self-service, improving the user experience of IT services
> 7. OS migrations (e.g., Windows X)
> 8. Application upgrades
> 9. Remote working initiatives (mobility, tablets, kiosk, remote connectivity)
> 10. Outsourcing/strategic partners
>
> To a greater or lesser degree, good ITAM data underpins or supports all these projects through faster decision-making, less cost, and fewer errors.
>
> We've learned that Stella's company is acquisitive. It's doing lots of M&A activity and is looking to drive economies of scale—to take two complementary companies and underpin them with a smaller combined network. Any projects, communications, or initiatives that Stella produces that are aligned with acquisition, therefore, will obviously be well received.

GUERRILLA MARKETING AND YOUR ITAM PRACTICE

Stakeholders who make use of your valuable ITAM data can become evangelists who promote the benefits of your practice to the company.

During a recent poll, only 38% of participants said they had someone outside their ITAM department who would act as an evangelist for ITAM.

Evangelists should be actively encouraged. You're looking for somebody who gets what you do and understands the value of ITAM. Perhaps they've had direct experience with it—they understand the value of your team and sing your praises to other people. This is communications by stealth; I've mentioned how valuable communications are to ITAM; evangelists are a great addition to your communications plan.

How do you recruit or encourage evangelists? Well first, you need to do good work. You need to educate people about what ITAM is all about, and you need to listen to people across the business: What are their challenges? What do they need help with?

You need to be seen as someone who acts on feedback; someone who actually listens to what people are telling you. I can't emphasise enough the value of communications. If you're doing good stuff in your ITAM department and communicating well, you'll undoubtedly recruit evangelists.

SUMMARY

We've covered how to identify and engage with stakeholders. Remember to think about the following:

- How are you hitting goals?
- What is your CIO up to?
- What is the company up to?
- What are the rest of the IT teams up to?
- How are your goals dependent on other people, and how can you help them so they can help you hit your goals?

Section 2: Take-off

Chapter 5 - Entitlement
Chapter 6 - Inventory
Chapter 7 - Reporting

Chapter 5

Entitlement

INTRODUCTION

How do we make software assets work hard for us? How do we generate every last drop of value from them? In this chapter we'll cover how to prepare for major contract renewals and negotiations, how to sweat software assets, and how to prevent overload with complex licensing.

WHAT IS ENTITLEMENT?

In a nutshell, entitlement is having the right to something. Entitlement is your right to use or take advantage of something. In the case of software, this means the right to use the software (the software license), and the terms and conditions under which this right is granted to you (the software licensing contract).

Commonly used in the ITAM industry is the term "effective licence position" (ELP). This basically means taking all of the things that you're entitled to and comparing them to how you are actually using them to determine whether you are compliance with the terms and conditions of the software licensing contract.

An ELP provides guidance regarding whether you are compliant with the terms and conditions of the licensing contract, and can also tell you when you may have purchased too many licenses, or may not be taking full advantage of the licenses you do own.

WHAT ARE PRODUCT USE RIGHTS?

Product use rights are the terms under which applications can be used. In the example below, the terms are specifically for desktop operating systems and grant special usage above and beyond the terms of an ordinary licence.

Desktop Operating Systems

Device License
1. Customer may install one copy of the software on a Licensed Device or within a local virtual hardware system on a Licensed Device for each License it acquires.
2. Customer may use the software on up to two processors.
3. Local use is permitted for any user.
4. Remote use is permitted for the Primary User of the Licensed Device and for any other user from another Licensed Device or a Windows VDA Licensed Device.
5. Only one user may access and use the software at a time.
6. Customer may connect up to 20 devices to the Licensed Device for file sharing, printing, Internet Information Services, Internet Connection Sharing or telephony services.
7. An unlimited number of connections are allowed for KMS activation or similar technology.

Source: Microsoft Volume Licensing Webpage

As you can see from Microsoft's list, managing entitlement is not just about counting licences and computers. There are extended use rights available to you for each piece of software. For example:

Point number 4 refers to 'remote use': "remote use is permitted for the primary user of the licensed device and for any other user from another Licensed Device or a Windows VDA Licensed Device" or Point 5 – "Customer may connect up to 20 devices".

There is versatility in the Microsoft terms to exploit. And there's a reason Microsoft has put these use rights in the contract—presumably because they will delight customers or provide competitive advantage. One of the ways of squeezing best value out of software is to use all of these rights to your advantage.

COMMON PRODUCT USE RIGHTS

Some of the most common product use rights include the following:

- **Upgrade and downgrade rights** – If, for example, you have a current maintenance contract or subscription, vendors will typically allow you to upgrade to the latest version. Similarly, some vendors will allow you downgrade rights (e.g., if you're on version 10, you're allowed to use versions 9 and 8, etc.).
- **Secondary use rights** – With these rights, you can use the application you've bought for your PC, for example, on a second device, such as a mobile device or a tablet.
- **Multiple versions on the same device** – It might be perfectly legitimate, according to the vendor, for you to have versions 6, 7, and 8 of a product, for example, on a device.
- **Virtualisation** – You might be allowed to virtualise the software, or stream it or some other way of taking advantage of the software in that fashion.
- **Connectivity** – Often there are different ways that you're allowed to connect to the software (through printing or sharing documents, for example).

Stella: Product use rights optimisation in action

Lisa is the licensing specialist at Stella's company, and she has two editions of the same application that she wishes to optimise: Factory Standard and Factory Professional.

Factory Standard costs $1000 per install and Factory Professional costs $1500, as per the table below.

		$1,000.00		$1,500.00
	Factory Standard	Costs	Factory Professional	Costs
Licensing	1,000	$1,000,000	750	$1,125,000
Inventory	2,000	$2,000,000	500	$750,000
License Balance	-1,000	-$1,000,000	250	$375,000
Factory Professional Downgrade Rights	-750	-$750,000		
Saving by applying product use rights	250	$250,000.00		
Not Used	40%	$800,000	60%	$450,000
License Balance (ELP and Usage)	50	$50,000.00	300	$450,000.00

Lisa researches the procurement history and finds she has a total of 1000 licences for Factory Standard. That's an investment of $1 million. She has 750 copies of Professional, which totals $1.1 million.

Lisa then takes a look at inventory data (covered in Chapter Six – Inventory). Reviewing inventory reports, she can see the company have 2,000 copies of Factory Standard installed ($2 million), and 500 copies of Factory Professional ($750,000). The licensing balance, at face value, is a compliance exposure of $1 million and a surplus of 250 licences. There's a surplus of 250 Factory Professional licences, which results in $375,000 in total wasted software. This licensing position is fairly typical of a corporate environment, a mixture of both compliance issues and overages.

Lisa now wants to take advantage of the product use rights for Factory to improve her licensing position.

Lisa knows that Factory Professional has downgrade rights—organisations are allowed to take the 250 surplus of Professional to counter the deficit of Standard, giving her a new balance of 750. Applying product use rights in this way saves the company $250,000. This is a simple example of the power of product use rights in action.

Now Lisa can take things a step further and look at how Standard and Professional are actually being used. She looks at her inventory tool which is tracking usage and can see that 40% of copies of Standard are not being used ($800,000 in potential waste) and 60% of Professional are not being used ($450,000 in potential waste).

The total licence balance, once product use rights and usage have been applied, has transformed from a deficit, a compliance hole of $1 million, to a surplus of $50,000, or $450,000 in the case of Professional.

In some instances, you might not be allowed to remove software, or there might be other restrictions, but even if you're not allowed to do any adjustments, knowledge of these numbers are fantastic bargaining chips when it comes to negotiation and renewal. Again, this is a simple example of how you can make best use of your entitlement and squeeze best value out of a product by taking advantage of product use rights.

FOUR STAKEHOLDERS AT THE TABLE

When it comes to negotiating with software publishers and suppliers, it's recommended that four key stakeholders sit at the table: Demand, Standards, Procurement, and ITAM.

There are caveats, of course. This won't apply to every organisation, and if it applies, it will depend on whom you're negotiating with, how much is at stake, and what else is going on at the company, but generally, each of these four stakeholders should be represented in some form throughout the IT contract life cycle.

Demand

Our first stakeholder in the lifecycle of a contract is Demand, which is basically "the business". Who is demanding this software, these services? Who in the business is making use of this software, hardware or services? If it's a data-centre renewal, it might be a system owner. It might be a representation of the end users. It might be a business unit. Whoever represents the demand of the business ought to be at the table to express their views.

Standards

The next stakeholder is Standards. Who is assessing the standards in terms of what software or devices are in use throughout the organisation? It might be the enterprise architect team or some other body within the organisation that determines the long-term strategy in terms of the software and services and systems being delivered. How does this contract fit in with that and how can it be shaped to fit those standards?

Procurement

I speak to a lot of organisations in the ITAM space, and I've found that sometimes this stakeholder doesn't get involved—sometimes it delegates to ITAM. But IT procurement should be at the table in some fashion throughout the life cycle.

ITAM

The final stakeholder is the ITAM team—you. This stakeholder provides the intelligence. It provides the evidence. It provides what's being used. It provides a snapshot of the entitlement. It provides options to all the other stakeholders, explaining, "This is what's actually going on the ground in terms of these assets"

If you're looking at your top ten software publishers and your strategy, in terms of best practice, these four stakeholders should be at the table.

Renewals and risks on the horizon

Lisa has developed a calendar of renewals, key dates, and potential risks for the next 18 months—a rolling calendar of events to monitor and prepare for. She can see the priorities and milestones in renewal cycles. She can also see who is doing what within the renewal cycles, and what the team perhaps won't cover due to limited bandwidth. Armed with this knowledge, Lisa engages regularly with Demand, Procurement, and Standards to ensure all stakeholders are represented.

LICENSING CLARITY

As mentioned in Chapter 1 – Authority, addressing the threat of software publisher audits is one of the biggest drivers for organisations introducing an ITAM practice. Two of the primary ways to reduce the risk of audit is to articulate to decision makers how licensing is measured in the contract and communicate the importance of measuring licence compliance on a regular basis.

> "License metrics, product use rights, and how to measure consumption should either be public domain or detailed in the contract." ~ The Campaign for Clear Licensing [4]

To prevent any ambiguity during audits, in terms of professional conduct and forecasting what you might be spending, it is recommended that you nail down exactly what you're going to be measuring in each one of your contracts.

Software licensing is an unregulated industry, and it's often difficult to tell the difference between genuine infringements of intellectual property and overzealous software sales reps. It sounds a bit draconian, but it's useful to determine exactly how you're measuring this product or these product suites or this whole estate, what you're entitled to, and how you're going to measure that consumption.

> John and Lisa have gone as far as to detail the exact executable files that need to be counted in their contracts with their top ten publishers so there is absolutely no ambiguity.
>
> John and Lisa are anxious to reduce some of the ambiguity around contracts to try to and preventing themselves being overloaded by licensing complexity. When speaking with key suppliers, John has insisted that the contract be measurable and has requested amendments to a contract to improve clarity regarding license terms and conditions and how they should be interpreted.
>
> In the first chapter, Stella put her business plan together and was talking about maintaining an ELP and audit-ready status for her top ten vendors. What is audit ready?

WHAT IS AUDIT READY?

Being audit ready is not about being 100% compliant at all times. That's not realistic (nor is it a good use of your time or resources). Being audit ready is about being prepared

---

[4] www.clearlicensing.org

so that if you do get an audit request, you know the drill. You have a playbook for what to do. You know whom to speak to, who the key players are, what you're going to do, and how you're going to respond. Although it might not be welcome, it won't be an onerous task.

And as your ITAM practice matures, if you've got good data, you'll be able swat away audit requests—auditors and software vendors will leave you alone. That's the goal. It bears repeating: this is not about being 100% compliant all the time. It's about just being responsive and knowing what to do.

By maintaining audit-ready status for your top ten publishers, you'll save a lot of time (people won't be distracted from key projects to defend audits), experience fewer surprises, and save a lot of money.

MANAGING VENDOR LICENCE STATEMENTS

A vendor statement, nowadays often accessible via an online portal, is the software publisher's view of your licence purchase history. If you contact your software publisher and ask, "What have we bought? Can you show me a statement of what we purchased in the past?" they will show you their version of history.

The vendor statement should be considered just one perspective. It might be inaccurate for several reasons, and it's your role as an ITAM professional to verify the accuracy on an ongoing basis.

COMMON ISSUES WITH VENDOR LICENCE STATEMENTS

- Missing agreements, subsidiaries, legal entities – You might log into your portal or request your licence statement from the publisher and discover that agreements

are missing, or that certain subsidiaries or sister companies aren't included.
- Missing licences, transfers, merger and acquisition activity, or concessions – Missing purchases are common. Or perhaps you've transferred something internally and this hasn't been recognised. Maybe you've bought or divested a company and you're still seeing that on your statement.
- Poor purchasing practices – although a requisition or purchase order was raised, no one checked to confirm that an invoice was received and paid so the license was never actually purchased.

As a recommended best practice, continually validate your license statement and help the vendor, using your evidence, get a more and more accurate picture over time. Unfortunately, this isn't a one-off exercise. When you place an order with a software publisher, the order is complete only when you can actually see it in the online portal—because that means it's been through the channel. It's gone through distribution, it's gone through the reseller, it's hit the online portal, and it's now recognised as a purchase by the vendor.

Don't be dependent on your reseller or vendor entitlement records and license statements. Accidents happen—agreements can get overlooked. Keep your own version of the truth so that you can validate other sources. Just like inventory, entitlement data needs to be cleaned and verified regularly for accuracy.

A final piece of advice regarding working with vendor licence statements: centralise and consolidate your account with publishers where possible. They appreciate this as well. For example, you could restructure, rename, or link your subsidiaries so that they are easier to find and not overlooked the next time a search is done.

For more on this topic, see the following discussion on managing vendor licence statements:

http://forum.itassetmanagement.net/2146881/Managing-vendor-statements-online-portals

SWEATING ASSETS

To sweat an asset is to extract more value, use, or output from it beyond its intended value. So what are some ways of doing this?

- **Resale/transfer** – The obvious way to sweat software and hardware is resale or transfer. If you're in Europe, you're allowed to resell perpetual software under certain terms. And of course, you can resell or transfer hardware. Some vendors actively encourage it and will facilitate transfer.
- **Alternative use** – Another way to sweat assets is to offer an alternative use. For example, I've seen company laptops once used for corporate sales later used in clusters in the data centre, and PCs turned into kiosks. You can extend the useful life of devices.
- **Extended use** – For example, if you have a three-year refresh cycle or a three-year business plan with ROI, you might extend that to four years.
- **Full exploitation of usage rights** – Look at the full product use rights for every product that you want to exploit and think of other ways to use that software and extend its use.
- **Increased knowledge of capabilities** – You can get much more value out of a piece of software by simply training people on how to use it properly—make sure they get full use out of it and they'll probably use it for a longer period of time.

- **Alternative delivery mechanism** – If you have an old copy of Adobe, for example, if terms allow, put it in a container or in some sort of virtual lockdown environment whereby you're able to use it forever. You don't have to go to Adobe Cloud; just lock down an old version. Virtualise it and ruggedise it so that it's protected in a corporate environment. Think about alternative use within the licence terms.

I've also seen people take an expensive piece of software and stick it on an old laptop. Rather than virtualising it and paying for the virtualisation fee, they literally hand the old laptop around like it's a USB stick. It sounds bit crude, but if you're cash strapped and looking to be creative with your budget and make best use of software, it's a viable alternative, and it's (most of the time, check the terms) within your licence rights. Again, it's about maximising your product use rights and fully exploiting them.

NEGOTIATION 101

> *"Give me six hours to chop down a tree and I will spend the first four sharpening the axe."* ~ Abraham Lincoln

Lincoln's quote on preparation is apt in the context of software contract negotiation. Ideally, to prepare for contract renewal, ensure the following:

- you have an accurate, trustworthy view of what you have;
- you know what you're using and what you're not using;
- you know where you want to go (with that application, that vendor, your IT strategy etc.)
- you have an idea of where the supplier is going and what their motivations are;

- you have input from the four stakeholders mentioned earlier in the chapter;
- you are early—you are being proactive; and
- you have access to all previous agreements and concessions in writing.

In this chapter we've covered the concepts of entitlement, product use rights and audit ready. We've discussed whom to involve in the renewals and purchasing process and how to negotiate and sweat assets. In the next chapter we'll look at the more technical side of things when we look at inventory.

Chapter 6

Inventory

INTRODUCTION

In the previous chapter we looked at managing our entitlement: what we've purchased and are entitled to use. In this chapter we look at inventory: what we are actually consuming. We'll look at how to build a technology strategy so you can monitor your environment, and how to build inventory data that everyone trusts.

NUMBER OF ASSETS: THE MOST BASIC OF METRICS

An accurate, up-to-date, and dynamic view of the total number of IT hardware assets in an organisation is probably the most basic of ITAM metrics to monitor. It seems like a simple number to track, but I'm constantly amazed by the number of organisations that don't have a clear view of it.

Tracking the total number of assets in an organisation may seem like a pointless goal, but it's from this number that many contracts, agreements, and structures are put in place. For example, it's not uncommon for an antivirus contract to be signed based on the total number of PCs, laptops, and servers in a business. Or this number might define the number of staff working on the help desk. In some instances, a Microsoft licensing agreement will be negotiated, agreed, and signed based on the number of IT assets in an organisation.

All of this leads to organisations' spending budgets based on a metric that might be wholly inaccurate. For example, in the last month, a large European brand name discovered a 25% deviance in the number of IT assets in their 4,000-asset estate. This is very common and could lead to a seven-figure discrepancy in their Microsoft licensing agreement alone—either a hefty overspend or huge compliance risk.

An IT asset count also serves as a basic litmus test to assess an organisation's ITAM maturity. Asking five different IT decision-makers in a business how many IT assets exist quickly highlights the degree of control in place.

When used efficiently, this is the most primary function of IT inventory tools and network discovery tools. It gives IT managers a clear, global view of all IT assets. Most important is the ability to show inputs and outputs in this global view, devices being added and removed from the network, so that decision-makers can make informed decisions based on up-to-date and accurate information.

Let's say your company owns 10,000 assets and your CIO says, "Prove it". What measures do you have in place to determine accuracy? In a ITAM Review recent poll, 47.2% of respondents claimed to be measuring accuracy, and 53.7% either weren't measuring accuracy or weren't sure if they were. Through this basic metric, many are leaving themselves open to potential exposure.

DECIDING ON A TECHNOLOGY STRATEGY

What tools do we need to track our consumption of assets and their status? What is best on the market? Which tool covers the most platforms? What key performance indicators should we measure, and what does everyone else measure? These are perhaps valuable questions, but they are very much bottom-up. Given the enormity of tracking huge technology environments, it pays to take a top-down approach.

> I'd like to introduce you to Petra, our ITAM analyst. She works for Stella and has been tasked with reviewing the company's approach to inventory. We learned that inventory and asset management generally is a bit poor at Eurolager. We'll follow Petra on her journey as she helps Stella. First, she's going to tackle technology strategy.

Top-down technology strategy

So what is a top-down approach? I recommend these three steps when assessing technology strategy:

1. First, look at the vision. What is the long-term goal of the ITAM practice? What is the high-level vision?
2. Second, determine what priorities you need to focus on to deliver against that vision.
3. Finally, look at entitlement.

If you asked me, "What's the best inventory tool for me?" I wouldn't look at the market, I wouldn't look at what everyone else was doing—I would look at your entitlement position. That might sound counterintuitive, but I'm going to walk you through exactly why this approach is important.

> Stella shared her vision with us in Chapter 1 – Authority. "The ITAM department will provide inventory of all assets with 95% accuracy, and an audit-ready status will be maintained for the top ten strategic software publishers with 97% accuracy."
>
> This is the vision she created to win senior management approval and get real influence behind her ITAM practice. She was suffering a lot of audits and the team's reputation was suffering from poor data. Remember, this is Stella's vision, not yours. Stella's plans ignore many of the benefits of ITAM to focus on something very specific that has been hurting her company. To learn how to build a strong vision specific to your company, refer to Chapter 1.
>
> This is the vision that allowed her to secure funding—the company stands behind this vision. Now it's time to move to step two. What are the priorities she needs to focus on to deliver against that vision?

THE 80/20 RULE

In terms of establishing priorities, we need to look at the good old Pareto[5] principle: the 80/20 rule.

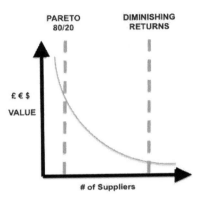

If I lined up all of your suppliers, from largest to smallest (left to right), you would likely see a red curve similar to that in the diagram above. Typically, you've got high-value suppliers, the big guys, at the left, and then a long tail of much smaller suppliers. Your list might run into the hundreds or thousands. The 80/20 rule suggests that 80% of your cost and risk is tied up in just 20% of your suppliers. Knowing this, you can significantly reduce the enormity of the task at hand and focus your efforts.

It's not that the rest of the suppliers aren't important. It's just that I've never met an ITAM person with too many people on their team. Everyone is stretched. You need to focus your priorities if you're going to deliver. Also, the good practices implemented for the top-20 will positively affect all suppliers.

---

5   Vilfredo Pareto (1848–1923) was an Italian economist who discovered, amongst many other things, that 80% of the land in Italy was owned by 20% of the population.

The law of diminishing returns also needs to be considered when managing your suppliers. At a certain point down the supplier curve, you're going to reach a stage whereby the cost of managing the supplier is more expensive than the asset itself. As asset managers with enormous estates to manage, we need to be cognizant of both the 80/20 rule and the rule of diminishing returns in order to use our time and resources wisely.

So we know the vision, and we know the high-value targets that will deliver our vision; the third and final step in deciding our approach to technology strategy is to identify the metrics required to measure consumption and progress with those high-value suppliers. If, for example, Oracle is one of my priority vendors and the licence metric is per processor, what data am I going to collect? What tools am I going to need in my data centre to measure a processor to be able to satisfy that metric?

A FOCUSED SHOPPING LIST FOR TOOLS

A top down approach to Technology Strategy allows you to provide a focused shopping list of requirements. Only after you've been through this process can you say with certainty what you need in terms of an inventory tools, or any other ITAM technology for that matter.

Tool types

The topic of this chapter is inventory, but it's worth pausing to reflect on the four main types of tools within the ITAM space:

1. Auto-discovery
2. Inventory tools
3. Normalisation and software catalogues
4. Software usage tools

Every large company already owns a handful of tools that may be helpful for your ITAM goals. Your company likely owns some sort of systems management, inventory, or software delivery platform. For example, the market leader is Microsoft system centre configuration manager (SCCM). You might have Altiris, BMC, or any number of other tools.

The primary purpose of software delivery tools is shipping software, pushing software down to desktops, and applying patches and upgrades. Typically, you will put an agent on each device, and that device will talk back to a central database. The output from one of these tools is a patchy inventory of what devices you own from a very technical perspective—it's usually raw data that is difficult to make sense of at first, but it's a start.

Auto-discovery

The role of auto-discovery is to find the devices that are not in the inventory—the ones that have been missed. This is really important to consider when you're aiming for complete coverage and accurate data. If you want to get data that everyone trusts, you need the safety net that says show me the devices that aren't being audited but then show me the devices that have been missed.

Auto-discovery is like a neighbourhood watch. It says, "Hang on, there's a machine over there on somebody's desktop, or there's a server in the data centre which isn't being inventoried regularly and we need to pick it up."

Tools such as SCCM might have data inside them useful to ITAM but shouldn't be considered ITAM tools. They don't typically include auto-discovery.

Software usage

The ITAM business plans for many organizations have the focus of removing unused software so they can reduce spend. You have got to think about how my going to do that? What inventory tool am I going to use to pick up consumption in order to identify unused assets? Tools such as SCCM are unlikely to do it elegantly. You need to think about what complements these kinds of tools. You also need to think about the scope of your assets.

Based on the priorities you identified as part of your technology strategy, you need to think about the assets that are important to you in regards to the high-value vendors you're looking to track. Are your most important assets in the data centre? The virtual environments? The desktop? On mobile? It's tempting to be pulled along by the sales spiel of the inventory tool supplier or the SAM tool supplier who says they cover every single platform. But what's right for you? What suits your environment and your goals? Make a decision from there.

Agents

The final consideration when building a list of tools we need is agents. Obviously tools like SCCM use an agent. There are also agentless inventory tools. Sometimes you might want to put agentless tools in the data centre and agents on the desktop, and so on and so forth.

My point is that whenever an asset management practice is doing really well, it's rare that they've got only one tool. When a practice has specific priorities, it needs a blend of tools to help them meet their goals. You need to consider all of these technology elements when looking at inventory and discovery tools.

WHY MEASURE THE ACCURACY OF INVENTORY?

If you're measuring the quality of your inventory data, you're going to be much more credible as an asset management practice. You're going to build up a great reputation if you have trustworthy data.

ITAM departments without good data are easy to bypass and ignore. They are seen as bureaucratic; as blockers. People wonder what's happening in these departments. But by having great data and providing great value, you bring people to the party. They want to get involved in projects because they see the real value ITAM offering.

Measuring inventory accuracy is also important in terms of your company's risk tolerance. For example, if you have 10,000 assets and you're looking at 95% accuracy, that's 500 devices that are unaccounted for. That might be 500 servers or 500 desktops or laptops. Consider the cost of replacing those 500 devices, including software support, software and licensing, and maintenance. What level of risk is your company happy to accept in terms of accuracy? If it's worse than 95% these numbers really start to creep up.

It's also possible that legislation will dictate that you must have accurate inventory—perhaps Sarbanes-Oxley or HIPAA. It might require you to produce accurate, transparent records with an audit trail to show your due diligence.

Finally, an accurate measure of inventory might be an audit requirement (internal audit or external) or a security initiative. For example, the standard Microsoft Products and Services Agreement states that if unlicensed use is 5% or more (i.e., your licensing is less than 95% accurate), Microsoft will charge you 125% of the current list price within 30 days. Obviously this agreement is for Microsoft software, but if you haven't got 95% accuracy on hardware, no way are you going to get accuracy on software. Both go hand in hand.

Most companies starting out will have no where near 95% accurate inventory, so they need to ensure that they identify machines that do not have an inventory tool installed, or where the agent has stopped working. This is another example of where working with stakeholders is critical – tools which aren't working correctly will often require a technical fix that ITAM professionals are not qualified to carry out, or don't have the appropriate access rights.

So now that you know why you should measure accuracy, how do you actually go about doing it? Let's see how Petra, our ITAM analyst, has been getting on.

> Petra has been investigating how to measure accuracy within her environment. She knows that previous audits are great sources of knowledge when looking at improving ITAM. In Petra's experience, auditors (taking either an internal or external approach) are looking for two things: coverage and accuracy. "Have you accounted for absolutely everything in your environment?" "If you've given me a number of 1000, is that number accurate and up to date? Have you actually got 1200?" "Have you got great visibility?"
>
> Petra has decided to put three practices in place to measure accuracy: physical spot checks; life cycle checks; and comparisons with other data sources.
>
> Physical spot checks are just as they sound—they involve physically eyeballing devices. As an asset manager, it's easy to lose touch with the actual physical devices when you don't see them often, when they're just records in your database. For this reason, it's valuable to do the occasional spot check, to physically look at the device and see if it's being reported accurately in your asset systems and Active Directory and so on and so forth. Obviously, some of you reading this work in multinational, multicountry organisations across hundreds of sites. This isn't a greatly scalable exercise, but it's a great way of sanity checking and building confidence in the fact that your data is accurate.
>
> Next is the life cycle check. An asset goes through transitions in its life cycle. It might hit the service desk for an upgrade, or get a new operating system through a migration, or get retired or transferred. For every one of those steps, you want to have a quick process in place, something that confirms that the asset data is actually reporting what is on the machine. Encourage people in the company, on the service desk, for example, to report any inaccuracies. Cleaning up the data and any inaccuracies will help them too, and increase their confidence in the data, not to mention improve relationships.
>
> Finally, Petra is going to compare her core inventory data with three data sources: inventory, Active Directory, and SCCM (there are many other sources you can use). Through these comparisons, she can find exceptions and see where things are missing.

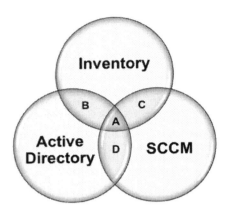

> In an ideal world, all of Petra's assets would be in section A in the diagram above; i.e., they would be in Active Directory, in SCCM, and would have an inventory client (A piece of software that monitors what is installed, how a device is configured and when changes are made to it)
>
> However, you might, for example, have assets in B, meaning they're in your inventory and they've got a record in Active Directory but they're missing from SCCM and shouldn't be. By gradually working through all of these data sets and managing the exceptions and acting on the data, over time, you can end up with incredibly sharp and accurate data that everyone trusts. Active Directory gets better, SCCM gets better, and people start feeling confident in your ITAM practice.

SOFTWARE RECOGNITION

Software recognition is another type of technology to consider when looking for help with inventory. It's sometimes called normalisation because it takes variants and normalises them to create common terminology. For example, your raw inventory data might have entries such as Microsoft Limited, Microsoft Software, Microsoft Corp., etc. The role of software recognition technology is to create a single entry—"Microsoft"—that everyone can work with.

The role of ITAM involves working with four "data languages", so in many ways, you are a translator and mediator:

1. Installation or consumption data, which is techie stuff
2. Product data, which is marketing stuff
3. Purchasing data, which is procurement stuff
4. Contract data, which is legal stuff

Let's say I buy Microsoft Project 2013—that's what the publisher calls it. However, what shows up on my network varies wildly. I look at the raw data provided by my inventory tool and find many files and technical junk that make up an install of Project 2013. Looking at the paper trail, the invoice, the procurement data, might simply say "Software", while the contract and associated product use

> Petra has been looking at her inventory data and realises it's a bit of a mess. She's been exploring software catalogues, looking at SAM tools to help her enrich the inventory she already has. Her goal is trustworthy inventory data that everyone understands. Because of the company's acquisitions, she operates in multiple countries and multiple territories, and because she's got different brands in different departments, she's got six different inventory sources. This is fairly typical—a mixed bunch of tools from different sources.
>
> It's not because a silver bullet of a tool doesn't exist—sometimes you might find a tool that does everything. It's because it's not always about the technology. Sometimes politics get in the way. Some office in some country wants to do it their own way, so you've got to deal with their data, their tool. Sometimes it's about dealing with what you have because it's all you have. Petra's six different inventory sources are a big bunch of noise. She needs to be able to deliver actionable licence information. She justifies the business value of looking at these catalogues and SAM tools by means of sheer man hours at the rock face.

rights provide a lot of detail around how the application can be used.

In dealing with these four languages, our role is to find the value, find the signal in the noise. Software normalisation aids in this process and can be justified purely in time saved. We're turning technical configuration data into actionable licensing information.

If you manually picked through raw inventory information from SCCM, for example, you would be looking at months and months and months of work to find the detail you need. This is the value of catalogues and plug-ins that help inventory tools. You can also see the risks and cost savings a lot quicker if you filter out the noise. This market can be quite complicated—sometimes software catalogues exist within inventory tools; sometimes they're in SAM tools. You can buy catalogues as a managed service or as a subscription.

The ability to recognise, normalise and filter is key. Here are some examples of several features we see in software catalogues:

- The ability to recognise software – the correct name, the correct version, the correct manufacturer; whether it's in a suite, a bundle, a family, a package.
- The application's licensable status – of all this stuff, what stuff needs a commercial licence?
- Product use rights – can you use it in virtual environments? On mobile devices? Etc.
- Categories – for example, what software creates PDFs? If you've got 14 different versions of it, maybe you can consolidate?
- Average cost – some tools now allow you, without having any procurement data, to look at your inventory and immediately see what is high value, what is most risky, etc.

Management by exception

These are just some examples of features in modern software catalogues. Managing accurate inventory is not a point-and-click exercise. It's not a one-off, either. It needs to be an ongoing process. You build up a reputation over time. One of the ways to do this is to manage inventory exceptions.

If I've been doing inventory on a server month in month out and then suddenly I lose touch with that server, that's an exception. That's something I need to escalate, something I need to investigate—I need to write a ticket or do whatever is necessary to deal with that exception.

This needs to happen on two levels – firstly is correcting that particular exception so that the data is accurate. The second is to identify why the exception occurred in the first place and implement process improvements to help prevent it happening again. This iterative process is why managing by exception is so powerful – the more you do it, the more accurate your data becomes. The more accurate it is, the more it is used and the more your stakeholders will identify exceptions and discrepancies in the data that are then addressed by process improvements. It's a virtuous cycle of improvement that will drive your data accuracy to where you need it to be.

In Chapter 1 – Authority, Stella was building up the business case for why people should look at ITAM. She guaranteed momentum in her practice by setting up a monthly ITAM board—a chance for all of the stakeholders to get together, review the progress of the ITAM practice. This monthly board is a good opportunity to discuss any exceptions proactively and agree process and other improvements that will drive greater data accuracy.

Petra has been actively involved with the inventory side of things. She's provided a scorecard that reflects exactly how inventory is doing from her perspective. She presents it at the ITAM board meeting every month along with recommendations for improvement. Everyone can see the progress, and everyone can see what actions are required and who needs to carry them out - often this will be one of the stakeholder groups rather than the ITAM team itself, which is one reason why obtaining buy-in and sponsorship for actions at the monthly board meeting is important.

Consider the scorecard:

	JAN	FEB	MAR	APR	MAY	JUN	JUL	AUG	SEP
Total inventoried devices	22,005	21,074	20,468	20,976	20,626	20,353	19,842	19,240	18,669
In Active Directory but not in Inventory	1,544	2,863	3,115	2,688	2,758	2,758	2,779	2,804	2,867
Unaudited devices	756	599	599	599	595	588	574	557	81
AWOL (not seen in 90 days)	3,532	2,863	2,279	2,279	2,023	1,554	1,187	882	637
Total potential devices	27,836	27,398	26,460	26,541	26,002	25,253	24,381	23,482	22,253
Accurate inventory	22,005	21,074	20,468	20,976	20,626	20,353	19,842	19,240	18,669
Accuracy	79.05%	76.92%	77.35%	79.03%	79.32%	80.60%	81.38%	81.93%	83.89%

In January, Petra has a total of 22,000 inventoried devices. Over the course of nine months it goes down to 18,000. And in January, there are 1500 devices in Active Directory that aren't in inventory. Some investigation is required. Is there a good reason for a particular device to be in Active Directory and not have an inventory client?

Petra has 756 unaudited devices, which means they're missing from inventory or not responding. The AWOL ones are those that haven't been seen in 90 days and she's got 3,500 of those. So, in January, she has 27,000 potential devices, and 22,000 accurately recorded ones in her inventory, resulting in an inventory of 75%. Over time, it rises to 83% per cent. You might think that's fairly sluggish progress, but it's actually very good for a 20,000-seat organisation.

These things don't happen overnight, and most importantly, Petra is demonstrating accuracy and slowly removing risk. Her team is gradually moving towards the target of 95% accuracy. As well, don't just present your information—ask your stakeholders for help. Petra says to them, "I've got 637 missing devices. It used to be 3,500, but we're still not happy. I need your help over the next month, the next quarter, to do something about this." The more accurate your data, the easier the stakeholders' jobs will be.

"GOOD IT INVENTORY BEGINS WITH ENTITLEMENT": DANNY BEGG, HRG

Danny Begg is a SAM Technical Specialist and supporter of The ITAM Review. In this video Danny describes why it is important to start with entitlement. Watch the YouTube video of Danny here:

https://www.itassetmanagement.net/2016/03/22/good-it-inventory/

Key points:
- Assess technology strategy for inventory and discovery.
- Gather trustworthy inventory data.
- Manage exceptions.

In this chapter we've covered the importance of a top-down technology strategy to ensure your organization is collecting the right inventory information to satisfy your business plan. In the next chapter we cover the final section in the take-off phase to ensure a long term healthy ITAM practice; reporting.

Chapter 7

Reporting

INTRODUCTION

An airplane is travelling between two cities; on paper, the route is a straight line between two points on a map. In reality, the plane is never really on course. The pilot is constantly adjusting direction based on the environment. Through constant correction, the plane reaches its destination.

In your ITAM practice, you need to understand what's going on in your environment and change course accordingly. This is where the value of reporting comes in. Implementing a world-class ITAM practice doesn't happen overnight. By the time you've finished implementing the originally stated business plan, the world will have changed. Modern ITAM should be iterative, flexible, and responsive; reporting allows us to show our progress and plot our journey accordingly.

This chapter covers ITAM reporting and, in particular, how to use reporting to maintain momentum in the longer term.

HOW TO MAINTAIN MOMENTUM IN YOUR ITAM PRACTICE

Reporting is a way of demonstrating your strategy in action. It shows:

- progress against the business plan;
- trends and risks;
- the results of our efforts, our outputs;
- whether priorities are still important or whether they need to be rejigged; and
- the value of our practice—it justifies the existence of the ITAM practice.

For our reporting to have an impact, people need to actually consume and take action on our messaging. We don't want glossy automated reports being fired about in emails that sit unread in inboxes; we want action. We want to ensure that actions are taken on exceptions, that there is consensus on how to address trends, and that the accuracy and value of ITAM improves over time.

WHERE REPORTING FITS IN THE 12 BOX MODEL

Reporting might seem like a dry and dusty ITAM requirement, but it's critical in pulling together some of the other boxes in the framework:

- Reporting shows the AUTHORITY that you are delivering
- Reporting shows everyone that you are executing against your business PLAN
- Reporting keeps the TEAM motivated
- Reporting keeps STAKEHOLDERS focused and engaged

Reporting is also useful for your own sanity! We all know that some days, working in ITAM feels like treading water. We all have days where we take one step forward and two steps back. Hopefully your reporting will show that, in the longer term, you're making a dent in the universe, making a difference, and that your key indicators are moving in the right direction.

WHAT SHOULD YOU BE REPORTING ON?

One of the most common questions newcomers raise within the ITAM Review community is "What should I be measuring? What key performance indicators should I be tracking?"

This has been covered in other chapters, but it's important to revisit in terms of reporting. Ultimately, your reporting should reflect progress against your plan. If you don't know what to report on, you're probably suffering from a lack of vision and a business plan that isn't concrete. There are plenty of interesting ancillary metrics and reports we could discuss, and there are specific reports to create for stakeholders, but your core reports should demonstrate progress, or not, against your plan.

Stella's dashboard

Here's another reminder of Stella's vision: "The ITAM department will provide inventory of all assets with 95% accuracy, and an audit-ready status will be maintained for the top ten strategic software publishers with 97% accuracy."

Again, your vision will vary depending on your priorities.

Stella is supposed to show an audit-ready status for the top ten strategic software vendors with 97% accuracy. So, if she's reporting an ELP for each of the ten vendors, and she is showing the trends, she can demonstrate that whilst she might not have met the accuracy figure yet, things are moving in the right direction. We covered ELPs in more depth in Chapter 5 – Entitlement.

Stella could also report on inventory visibility, and highlight trends in this area. Some specific examples of this are covered later in the chapter. Again, she could show her senior management team and stakeholders that she's moving towards her target of 95% accuracy.

Finally, because her business plan is specifically aimed at software vendor audit risk, Stella could also keep a log of recent audit activity.

In summary, to demonstrate progress against her plan, Stella could generate three high-level reports:

1. An ELP for her top ten strategic vendors and trends
2. Inventory visibility and trends
3. An audit log

AN INTRODUCTION TO CSI

When it comes to reporting and maintaining momentum in your ITAM practice, it is useful to think about the concept of CSI—not crime scene investigation but *continual service improvement*. This probably isn't as exciting as examining blood on the carpet or DNA samples, but it is very important when it comes to building ITAM practices that last.

CSI is a term borrowed from IT service management (ITSM).

> "Continual service improvement, defined in the ITIL[6] continual service improvement volume, aims to align and realign IT services to changing business needs by identifying and implementing improvements to the IT services that support the business processes." ~ Wikipedia

So what does that mean for us in ITAM? Everyone working in ITAM is delivering services to their business. We can use CSI to ensure our services mature and improve as the business ebbs and flows.

WHY DO WE NEED CSI?

As mentioned, building a world-class ITAM practice doesn't happen overnight. In a large company, it might take you between two and three years. The goals that you set out on day one are going to be very different from the goals you end up with because your business environment will evolve. Your software and hardware and the services you use will evolve.

Your goals in the business plan are also going to change.

---

6   For more information on ITIL, visit www.axelos.com.

Everyone reading this book works in a business that is either growing or contracting, buying or divesting business units, and changing technology and ways of working.

These days, IT departments are looking at agile ways of working. They're looking at DevOps, at moving more quickly and being more responsive to customer requirements. As an ITAM practice you can't be stuck in the dark ages. You need to keep pace with the speed of change to stay relevant.

There's no point adhering to a business plan that you wrote three years ago if it hasn't kept pace with the way your business is evolving. The CSI approach means adapting and updating your business plan to make sure that you are continually relevant and continually providing value.

Stella's audit log

The Excel sheet below is Stella's audit log, which she uses to demonstrate historic audit activity. This is a tangible, concise report that shows the impact of her team's efforts.

Audit Log

	Publisher	Scope	Initial Notification	End	Status	Estimated Exposure	Demand from Vendor	Actual Settlement
1	BSA	Romania	Jan-13	Apr-13	Completed			
2	Oracle	Germany	Sep-12	Aug-13	Completed		$700,000	$700,000
3	Oracle	Subsidiary	Not Known	Oct-13	Completed		$150,000	$150,000
4	IBM	France	Sep-13	Jun-14	Completed		$13,200,000	$1,500,000
5	BSA	Australia	Aug-14	Jun-16	Completed			
6	BSA	Poland	Aug-14	--	Completed			
7	Autodesk	Global	Jan-15	--	Completed			
8	BSA	Subsidiary	Aug-15	--	In Process	$100,000		
9	Microsoft	Subsidiary	Oct-15	--	Completed			
10	Adobe	Subsidiary	Jun-16	--	In Process	$1,500,000		
11	Microsoft	Global	Oct-12	Nov-12	Thwarted			
12	Oracle	Subsidiary	Sep-12	Mar-13	Thwarted			
13	Oracle	Subsidiary	Nov-13	Dec-13	Thwarted			
14	Adobe	North America & Europe	Mar-14	Apr-14	Thwarted			
15	Microsoft	Netherlands	Oct-14	Oct-14	Thwarted			
16	IBM	Global	Mar-15	Apr-15	Thwarted			
17	SAP	Subsidiary	Oct-15	Nov-15	Thwarted			
							$14,050,000	$2,350,000

The first column indicates that Stella's company has experienced 17 audits in the last few years. This number in itself is a great reminder of why her ITAM practice exists, and of the continuous threat of audit.

The next column indicates the software publisher involved in the audit.

> Then there's the scope—Stella's company has a global presence, and some audits involved a specific country or business unit.
>
> Stella has also recorded the start and end dates of audits, which is useful in determining how much resource and management overhead is being consumed defending audits.
>
> As shown, a few audits were completed, a couple are in progress, and a few got "thwarted". Thwarted is really good—a real milestone for an ITAM practice. An audit is thwarted when you get approached but proceed to take the wind out of the vendor's sails straightaway with good data, saying, "Look, we don't even need to do an audit. Here's our data." This is the ultimate goal of an audit-defence team.
>
> Stella has also begun to populate the estimated exposure and financial penalties in her audit log. With these figures, she can truly justify her challenges. She can justify why she is doing audit defence for the company and the scale of the issues and risks involved. She can hopefully also demonstrate progress over time.

IMPLEMENTING A CSI PROCESS

OK, so much for CSI theory—how do we go about doing some of this stuff?

The Deming cycle[7] is commonly cited as a methodology to consider for CSI. In layman's terms: build a feedback loop and listen to it. The Deming cycle suggests four major steps.

- PLAN – Determine what is required in your ITAM practice to improve efficiency, effectiveness and financial impact.
- DO – Execute the plan as a project with a specific goal.
- CHECK – Verify success via measurable metrics or perhaps an audit to check accuracy.
- ACT – Take action on any findings or lessons learned.

Constantly repeat this cycle looking for ways to improve and align with business priorities. The goal is to steadily

---

7   Deming Cycle and Plan, Do, Check, Act: https://en.wikipedia.org/wiki/PDCA.

improve your ITAM practice over time in iterative steps, increasing the quality, accuracy, and impact of your efforts.

Many ITAM departments are quite reactive. They react to vendor audits or they react to their environments. To employ CSI in a Deming cycle is to be proactive. Take the bull by the horns and say, "Right, here's an area of improvement for our ITAM practice. We're going to break that off and focus on it. We're going to plan out what we're trying to achieve and then go away and deliver that as a project with specific outcomes. We're then going to check progress using metrics that we've agreed to at the planning stage. Then, we'll act on any findings as a result of that process."

The goal is to gradually improve quality. For an ITAM practice, that might mean cost savings; it might mean improving audit-defence times; it might mean cost avoidance; it might mean turning around an asset request in record time —whatever is important to you and your business.

Stella's CSI plan

Stella's original business plan, covered in Chapter 2 – Plan, referred to a terms of reference and ITAM board. She was able to say, "I've got the CIO's blessing to build this ITAM practice. These are the metrics I'm going to use to measure our success, and I'm going to go back to the CIO if anyone is deviating from our plan."

She then initiated monthly board meetings to get her stakeholders together to review progress, measure and review trends and metrics, and take action. This is how CSI is done in practice. I'm not saying this is the only way to do it, but this is the way that Stella has chosen to monitor her progress and be proactive.

Stella's inventory reporting

Another example of Stella's reporting can be found in Chapter 6 – Inventory. The table below is is another report that shows that not every month will be perfect: "We haven't got things perfectly nailed, but generally, the trend is going in the right direction." This type of reporting could be particularly useful for winning over sceptical stakeholders as you gradually increase the quality of your data.

The move from 79% to 83% accuracy isn't miraculous, but it's progress nonetheless. It's a trend in the right direction.

	JAN	FEB	MAR	APR	MAY	JUN	JUL	AUG	SEP
Total inventoried devices	22,005	21,074	20,468	20,976	20,626	20,353	19,842	19,240	18,669
In Active Directory but not in Inventory	1,544	2,863	3,115	2,688	2,758	2,758	2,779	2,804	2,867
Unaudited devices	756	599	599	599	595	588	574	557	81
AWOL (not seen in 90 days)	3,532	2,863	2,279	2,279	2,023	1,554	1,187	882	637
Total potential devices	27,836	27,398	26,460	26,541	26,002	25,253	24,381	23,482	22,253
Accurate inventory	22,005	21,074	20,468	20,976	20,626	20,353	19,842	19,240	18,669
Accuracy	79.05%	76.92%	77.35%	79.03%	79.32%	80.60%	81.38%	81.93%	83.89%

MAINTAINING MOMENTUM IN YOUR ITAM PRACTICE":
COLIN SIMMONS, KINGFISHER

Now that we've talked about CSI and have gone through what Stella is doing in terms of reporting to keep people on board, I want to share with you some feedback from a real practitioner out in the field. Colin Simmons is a nice guy, an experienced ITAM professional, winner of the ITAM Review excellence awards 2016 lifetime achievement award and a long-term supporter of the ITAM Review.

He has at least a couple of decades' experience in senior ITAM positions for large corporations in Europe and is now working as group IT asset manager for a company called Kingfisher, a holding group which owns some significant European brands. I asked Colin for his views on reporting and how to maintain momentum in your ITAM practice.

View the YouTube video of Colin here:

https://www.itassetmanagement.net/2016/09/27/colin-simmons-kingfisher-maintaining-momentum-itam-practice/

Key points:

- Expect your reporting requirements to change over time as your ITAM practice matures and its reputation is built.
- Scale back ITAM board meetings to once every couple of months or once a quarter after trends are established and a reputation is built.
- Ensure that business reporting for senior management is concise.

Reporting, a clear idea of our progress and trends completes the "take-off" phase of our ITAM implementation. In the next section we can turn our attention to keeping our ITAM practice airborne as a business as usual function.

Section 3: Cruising altitude

Chapter 8 - Transition
Chapter 9 - Request
Chapter 10 - Dependencies
Chapter 11 - Reclaim
Chapter 12 - Verification

Chapter 8

Transition

INTRODUCTION

Transition is an important component of the 12 Box Model because all organisations go through the process of change; environments are constantly changing.

Think of all the changes that have happened in your organisation in the last 12 months—from the minor upgrades to the major IT transformations. The root cause of every change can likely be grouped in one of six areas:

- New projects
- Requests for change
- Requests for new assets
- People (joining, leaving, changing roles)
- Engineers fixing things
- People buying things

	Source of change	Description	IT Function
1	New projects	Deployment of new projects or upgrades	EA, Projects, Development, Operations
2	Changes	Changes to infrastructure and production systems	Change Management
3	Requests	End user or IT team requests	Service Request
4	People	Joining, leaving, role change, finishing projects or otherwise not using assets	HR / Service Desk
5	Engineers	Engineer, support staff changes, IMAC activity	Incident Management
6	Buying	People buying stuff	Service Request

By addressing and embracing change, we can be proactive. We can be strategic with our approach to ITAM. If we don't learn to work with change, ITAM will always be a reactive function; we'll always be trying to identify what happened and clearing up the mess afterwards.

WHY BOTHER BEING PROACTIVE?

A few years ago, when IT projects took two years to deliver, you might not have been bothered about knowing about your compliance status with Microsoft or Adobe because you were on a three-year agreement cycle and changes were taking place more slowly. But modern IT projects are often smaller—they're often iterative and more responsive to customer requirements.

> Stella's company is interested in investing in a new service desk. In her business plan, she identified that working with a service desk was absolutely critical.
> In this chapter we'll look at three elements of Stella's journey: the opportunity to integrate ITAM and ITSM, the potential to work with other IT teams, and the potential to share the workload.
> 1. *ITAM and ITSM integration isn't about creating more work for both parties—it's about working smarter and reducing workload for both. We're going to look at how Stella achieves that.*

ITSM AS ENGINE ROOM OF IT

When it comes to IT changes, ITSM is the engine room. It often delivers change. But you need to consider that ITAM and ITSM have completely different motives. ITAM is predominantly about governance, risk (sharing risk, mitigating risk, addressing risk), and cost (cost efficiency, cost reduction, showback). ITSM, on the other hand, is ultimately motivated by service: "Let's get people back on their feet, let's make sure that service is being delivered, let's make sure we're up and available at all times, let's make sure that our services are being delivered as efficiently as possible, etc."Despite the different motives and priorities, ITAM and ITSM typically sit on the same data, which is why working together can be a great opportunity.

WHY SHOULD ITAM AND ITSM WORK TOGETHER?

Working with ITSM, ITAM can:

- meet compliance and efficiency goals whilst decisions and changes are being made rather than attempting to clean up a mess afterwards; and
- be productive and access the assets they need whilst being compliant.

Working with ITAM, ITSM can:

- offer visibility of asset relationships and costs to facilitate faster incident resolution, proactive problem management, and less error-prone change management; and
- deliver self-service without risks.

Let's say one of your end-user customers has an issue with their laptop. You're much more likely to be able to resolve the incident if you have up-to-date, reliable asset data. Simply by seeing the status or recent changes to the device, the operator can start resolving the incident without even speaking to anybody.

The service desk operator can then say, "Who else might have that configuration?" And you can be proactive and say, "Let's do problem management and look at why this is occurring and who else might have this issue as well." Then you can do less error-prone change management. If you're aware of the change management process, you can help the business avoid a bullet in terms of data centre errors that result in huge licence penalties.

ITAM can also help ITSM by delivering self-service without risks. Self-service portals and helping users to help themselves is very popular. Why wait on hold on the phone when you can log a request from the comfort of your pyjamas at home at 3am. Delivering a slick and seamless service to users is great, but the business still needs to meet its governance goals. You still need to be compliant. You can do that if you merge ITAM and ITSM. You can get the best of both worlds.

DATA CENTRE CHANGES: STAYING ONE STEP AHEAD

In a recent online poll we asked ITAM Review readers: "Does your ITAM function have visibility of changes being made in the data centre?" This visibility is typically in the form of the change management process.

As you can see from the table below, 51% are tracking changes in the data centre, 49% aren't, or aren't sure. There were slight differences between our respondents in different time zones.

Does your ITAM function have visibility of changes being made in the datacentre?

	UK	USA	All
Yes	54%	48%	51%
No	29%	42%	36%
Not Sure	17%	10%	14%

Based on 132 responses

Why is it important to track data centre changes? Significant costs and risks exist in the data centre. If you don't have visibility, you have to clean up the mess afterwards. By working with changes as they arise, you can help the business avoid nasty costs and risks.

> Stella has been working closely with the ITSM department. She's trying to embed ITAM's discipline into day-to-day operations. This is a really smart way of doing ITAM. It's not about creating more work; it's about just slightly correcting what people are doing to make sure that you hit your ITAM goals.
>
> She's found a number of opportunities in her initial investigations: the company is looking at a new service desk, they're exploring self-service, and there's an opportunity for a single system of record. By doing service management and ITAM together, you can actually update your ITAM repository in real time. Obviously it's never going to be perfectly accurate, but you're going to have a lot more accurate data than you would if you were trying to play catch-up.
>
> Stella has also recognised that she can de-risk the change management process. This isn't about bureaucracy. The last thing you want in change management is more red tape, but when you de-risk correctly, you add value and intelligence to the whole process. She's got a great opportunity to do this.
>
> She has also identified that working with the service desk is a great opportunity for a smoother and more secure HR process for leavers. Tracking company leavers is essential for hardware asset management. You don't want laptops with company data on them hidden under desks or left at home, even if the data has been encrypted. ITAM can work with HR and ITSM teams to claw back assets, hardware, and software and save a lot of money.
>
> Stella has also realised that quicker incident resolution is possible. In my experience, front-line service guys and gals like checklists and processes to allow them to close incidents as quickly as possible. If I'm going to rebuild a machine, I want use the same software and I don't want it to cost the earth. Checklists help make sure that errors are reduced, that the person with the issue is back on their feet as quickly as possible, and that Stella meets her ITAM goals.
>
> She can also help the enterprise architects realise their vision by adding commercial relevance and value to their long-term IT strategies. High-level blueprints can be supported by more accurate financial forecasts.
>
> Finally, she recognised that she can help reduce grey or shadow IT and the number of people avoiding the IT department altogether and buying their own equipment—she can help ITSM deliver a fast and efficient service.
>
> These are some of the immediate opportunities that Stella has identified.

"ITAM AND ITSM WORKING TOGETHER": GILLIAN LEICESTER, SYNYEGA

Gillian picked up the award for ITAM Professional of the Year in the ITAM Review 2015 Excellence Awards.

She has a wealth of experience at the rock face, and in terms of software waste, she has stripped hundreds of millions out of organisations. What does she recommend in terms of working with other IT teams?

See the YouTube video of Gillian here:
https://www.itassetmanagement.net/2016/02/16/how-to-harness-itam-and-itsm-integration/

I asked Gillian, "How do you work with broader IT teams? We don't want to work in isolation in ITAM—how can we work with these other IT teams so that we're a peer and we're not a nuisance, or a barrier? How can we work as if we're a true stakeholder?"

Key points:

- ITAM can't work in isolation but must consider IT stakeholders across the business.
- SAM can help de-risk and enhance projects or changes.
- SAM should get involved with the enterprise architecture community and help them, with commercial information, turn IT vision into reality in a cost-effective manner.
- A service request catalogue is key when it comes to integrating ITAM and ITSM; it allows requests to be delivered efficiently whilst maintaining ITAM accuracy and helping execute on a reuse-before-buy policy.

Let's look at how Stella is getting on with the service desk. She's got a new friend called James. He's the head of service delivery, and he's responsible for the new service desk. They've discovered specific ways to help each other.

First of all, Stella is going to sit on the change board. Change boards might be electronic, or they might take place in person. They vary in their sophistication and importance, but they ensure visibility, which is always a huge step. The IT asset manager brings a lot of visibility to the table in terms of what's going on in the company. This is incredibly valuable to the change management process. James will encourage Stella to join the change board, as the team will benefit from better visibility of assets, commercial relevance and how assets are configured.

They've also decided that Stella will get involved in the new self-service design, and they're in the process of choosing a service desk. Getting Stella involved from the get-go means she can embed her ITAM discipline into it (management approvals, checking for stock before purchase, etc.).

They've identified that HR have a batch process overnight that notifies people about the company's leavers and joiners. If Stella can refine the process to include ITAM, she'll have a great opportunity to get visibility on what's going on in terms of joiners and leavers. Being on top of the people leaving the company is a great way of clawing back assets as staff exit.

Stella and James are also going to review the incident management process as part of the new service desk, and Stella is going to be part of that process to help that team.

Finally, Stella is going to sit in on meetings with the enterprise architects team. Enterprise architects typically have long roadmaps. Long-term IT projects aren't dreamt up overnight. Stella's is getting involved in the long-term vision by adding IT asset intelligence will help this team realise this vision.

These are just some of the practical tactics that Stella and James have decided to act on in terms of ITAM and ITSM integration.

SUMMARY

There are great opportunities to integrate ITAM and ITSM. It's possible to work with other teams and add value to their day-to-day processes. It's about sharing the workload and collaborating; it's not about creating blockages and more work.

Chapter 9

Request

INTRODUCTION

In this chapter we address how to automate the asset-request process within your ITAM department via online asset-request systems. When a customer indicates that they want new equipment or services from the IT department, this is an asset request. Requests might range from a new mouse to a new data centre.

Enterprise application stores or online asset-request systems are win-win options. They present a great opportunity to automate ITAM and alleviate customer-facing admin whilst providing a better customer experience.

Best of all, we can break in our governance and policy requirements within the system so that our ITAM goals are automatically met.

This chapter will cover some of the reasons to consider an automated asset-request fulfilment system and features to consider when implementing one.

WHY CONSIDER AUTOMATED REQUEST FULFILMENT?

First of all, app stores are fast and efficient. With a smartphone, I can order assets at 3am in my pyjamas, I get immediate acknowledgement, and I don't need to speak to anyone to get the job done.

There are several reasons to consider automating the asset-request process:

1. **Less administration** – Once your company understands the benefits of ITAM, you'll undoubtedly start to get requests and enquiries. By automating and streamlining the asset-request process, IT asset managers can focus their energy on strategic objectives rather than getting bogged down in administration.
2. **Happy users** – Self-service and automated requests will speed up the experience for customers and increase their productivity.
3. **Automated ITAM governance** – ITAM governance goals can be designed and automated within a self-service arrangement.
4. **Service level agreements (SLAs)** – Automated request processes may facilitate faster overall responses. If you've never hit SLA targets, or if you've struggled to hit them, self-service can help you get there.
5. **Less diversity** – Fewer non-standard assets and less variation between assets (by steering users towards a specific menu of options) will mean that assets are easier to secure, support, licence, and manage.
6. **Internal charging** – Automated management approval facilitates departmental charging/showback.
7. **Automated re-harvesting** – Automated request systems can reverse the whole process to save money; re-harvest and reclaim as quickly as you deliver.

Stella and James are working together to create an asset-request process from scratch. It will be built into the new service desk. We will review their process and look at opportunities for automation along the way.

Stella's predominant focus is software, but James believes that many of the same principles can be applied to hardware requests.

The first thing that Stella and James discuss is where the asset-request process starts. How can it be initiated? What is the architecture they need and where are they going to direct users who wish to request an asset? They agree that the process will be best initiated within the support and self-service page; they don't want to confuse users with a different login page.

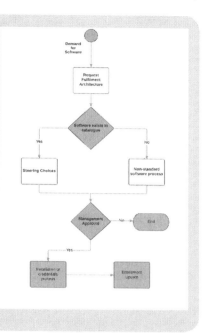

AMAZON AND ITAM

Once the end users have a destination, somewhere to go to request assets, we can steer them towards the right options. Consider building a preferred asset list, catalogue, or menu of options for users, to streamline the process.

E-commerce giant Amazon serves as inspiration on self-service best practice and steering user choices. At the heart of Amazon's success is the ability to provide a great experience for customers—one with as little friction as possible. Amazon customers can:

1. Search for an item and be presented with an algorithmically calculated list of options.
2. Read full descriptions of the item to understand exactly what they are purchasing.

3. Read ratings from customers to get feedback on the item.
4. See the different versions of the item (e.g., paperback, hardback, electronic, etc.)
5. Get a price.

You can use similar techniques to help your users make better software choices; provide a search facility, a clear description of each application, feedback from other users, details on the different editions available (standard vs. professional, for example), and pricing details. From an ITAM perspective, edition and price are particularly important; we want to ensure the user is picking the most efficient edition for their requirements, and the price reinforces that asset requests cost the company money.

OTHER STEERING OPTIONS

Reading customer reviews on an e-commerce store like Amazon or TripAdvisor can be useful when buying online, and customer ratings and other social feedback can also be useful in a corporate environment. Social feedback can steer your users to the right decision, help speed up their decision-making process, and also provide valuable feedback for contract negotiations. IT buyers can look at the summary of user feedback and see whether users are actually satisfied with the assets they're consuming.

Another thing to think about is building auto-reclaim or lease options into user choices. In doing so, you set the expectation: "I'm going to have this back when you're done with it, or I'm going to have this back if you're not using it." These days, many asset-request systems have the ability to add auto-reclaim options.

Where possible, you also want to direct people to cheaper or more appropriate choices. You may want to give them viewers rather than full editions. As mentioned, highlight the cost to the company. Make sure they know that this is why you're doing this. This is why you're doing ITAM. This is what ITAM is all about. It's costing the company money, and you need to emphasise the cost even if you're not doing internal charging.

A final thing to consider when steering choices is allowing users to help themselves. What we're seeing in modern SAM tools and shopping cart systems is what you would see in your iTunes or your Google Play account: what you've purchased, what you're using, what you can give back. Allow users to help the company.

WHY CARE ABOUT THE USER EXPERIENCE?

Why should we bother with the user experience? Why should we care in the ITAM department?

First of all, you're much more likely to give up assets, whether software or hardware, if you know you can get them back again. Say I'm a team leader and one of my team members has left and given their laptop back. I'm much more likely to give the laptop back to IT rather than stick it under my desk if I know there's a great process for getting it back again. People will give up assets not in use if the process is slick and they enjoy it. They're less likely to source elsewhere; they're less likely to bypass the IT department and look at shadow IT sources.

Requests, service, and support may be the only time your users actually contact IT. It is the shop window of the IT department. We need to make sure we're pulling our weight and doing the best we can for the entire IT department. A lot of companies that I speak to these days are trying to bring IT out of the back office in the basement and into the light as an enabler. A slick asset-request process is a great way to begin that transformation.

NON-STANDARD REQUESTS: GORDON RAMSAY VERSUS MCDONALD'S

Maybe you've got your preferred asset list and are steering users to make the right choices. But what happens if you've got requests for non-standard software or hardware?

Before we dig into the specifics, it's worth thinking about the culture in your company in terms of serving users. Consider two extremes: fine dining at Gordon Ramsay's Michelin-star restaurant versus grabbing a burger on the go from McDonald's. I'm not saying either one is right or wrong, and wouldn't dare offer an opinion on your choice of restaurant! These are just two radically different styles of customer service.

At a posh Gordon Ramsey restaurant, you're going to get expensive and personalised service. At McDonald's, whether you're in Beijing or London, you're going to have pretty much the same, efficient experience.

At the ITAM Review we've seen examples from both ends of the spectrum in terms of IT departments.

- Example 1: A financial services company. If traders on the trading floor don't get the support they want from IT, monitors and keyboards get flung across the department. The mantra here is "we must provide the user whatever they want and get out of the way".
- Example 2: A construction company with strict request options. Users are not allowed to deviate from the list, and the mantra is "like it or lump it". Perhaps they operate with a marginal profit base and have to be very, very efficient.

In terms of dealing with non-standard requests, you need to think about where you sit on this spectrum.

WHO DECIDES ON NON-STANDARD REQUESTS?

When you receive non-standard requests, which teams support you? I would hope the ITAM team is not working in isolation.

Security need to be aware of non-standard software requests to see what is being introduced to the network: is it part of your existing standards? Does someone in enterprise architecture need to review it? Can the service desk support it? New products and editions are a team decision rather than solely an ITAM one.

Requestors of non-standard assets should be made aware that their request might take a little longer to deliver. For example, it might take you two weeks to find the terms and conditions of the contract and to determine whether you're allowed to use it on the network.

The process and decision-making around asset request can be automated, but there is no set way of delivering it, as it is dependent on your culture and attitude towards requests. A standard list is a lot easier to support and secure but might lead to less flexibility and productivity. Will you bend over backwards for any request made, or direct users to a standardised menu?

MANAGEMENT APPROVAL

Once you've got a request for either a standard or a non-standard asset, you need to go through the management approval process. I would urge you to seek management approval regardless of whether you need it or not. It's a great way of providing accountability and stopping people from wasting time picking assets just because their colleagues are using them. Whilst we might be aiming for a frictionless transaction like Amazon, a little accountability doesn't hurt to avoid frivolous requests.

INSTALLATION AND CREDENTIALS

Then we need to look at deploying and provisioning of these assets. This is another great opportunity to automate; hopefully some of you are already doing this. Once you've got an approved software request, you can integrate it directly with deployment systems.

I've seen integrations with SCCM or Active Directory; if somebody is a member, they automatically get these assets.

Perhaps the future is about users and software as a service and cloud, and less about SCCM and Active Directory. Perhaps it's more about identity management systems or single sign-on or other credential systems. Cloud based assets are another great opportunity for automation in both deploying software and removing them in the same fashion as perpetual assets.

UPDATING ENTITLEMENT RECORDS

If possible, automate the updating of entitlement records. Hopefully you don't just push out software automatically. You want to be able to update or gate things depending on your entitlement.

This is the final stepping-stone of the process. The goal is to be able to automatically update entitlement without going into spreadsheets or updating records.

Just because you've got spare licences or assets doesn't mean you shouldn't charge or get manager approval for an asset request. Agreed management approval for an asset and existing entitlement should be two separate concepts.

Another consideration for entitlement is batching procurement; you don't need to kick off a set purchase-order process for every deployment. This would be expensive and inefficient. It's an expensive and inefficient way for the vendor to receive orders as well. Much better to batch things on a quarterly or even an annual basis. Many vendors are open to this.

There are specific product use rights to take advantage of here too. Some vendors allow trials, or state that the product is only licensable if used for 90 days, and so on. As long as it's all documented and agreed, most vendors will be happy to go along with your procurement process. It's also a great signal to the vendor that you've got your ITAM house in order.

For example, throughout the course of a quarter, I installed ten new copies of something but removed six because they weren't being used. Now I'm placing an order for only four instead of ten: this is the ultimate in automation and reduction in administration.

Asset request is a great opportunity to automate processes whilst baking in ITAM goals and governance rules and being responsive to customer requirements.

Some final policy considerations/approaches:

- Implement a reuse-before-buy policy (squeeze best value out of existing before buying); good asset-request systems should facilitate this.
- As mentioned, separate the concept of request from entitlement. Many asset-request systems can check licence availability on the fly—this is great but can be a bit of a lottery for the user. Why should one user get an asset for free because there happens to be a surplus at that time? Why not charge for all requests and accrue surpluses (continue to charge users when a surplus asset exists) as profit for the ITAM team?
- Remember that just because you bought a perpetual licence or annual subscription doesn't mean you have to provide this to the users—allow them to lease it for shorter periods, or request it back if it's not in use for a demonstrable length of time (e.g., 90 days).

- Consider tactical re-harvesting campaigns using your request system to remove high-value applications; good tools should facilitate this.
- Finally, consider your architecture for delivering the asset-request system. Users typically don't want many different systems for IT login—many organisations embed asset request into ITSM request fulfilment and incident management systems.

Implementation options to consider:

1. ITAM embedded within ITSM: ITAM tools can embed themselves into ITSM solutions so that whilst the user thinks they are using one pane of glass, they are actually using an ITAM tool embedded within the service desk.
2. ITAM integrated with ITSM: Alternatively, other ITAM tools provide the ITAM plumbing behind the scenes whilst the service desk itself executes all workflow requirements.
3. ITAM and ITSM: some ITSM toolsets includes ITAM as well.

In this chapter we've covered the request process and scope for automating ITAM processes where possible. In the next chapter we'll look at how to manage the many data sources that are required for successful ITAM.

Chapter 10

Dependencies

INTRODUCTION

This chapter covers the different data types we need in order to maintain compliance and how to conquer licensing. As licensing becomes more complex, it is even more important to manage both the data required to measure consumption and the dependencies between data sets. We will look at how critical it is to manage dependencies if you want to build a world-class ITAM practice.

HOW TO CONQUER COMPLEX LICENSING

Let's revisit some of the principles we've covered in previous chapters. The following four guiding principles will help us manage the complexity of modern licensing:

1. If you can't measure it, don't consume it
This is fairly self-explanatory: if you can't measure something, don't buy it. Enforce this principle among your procurement colleagues.

2. Get measurement metrics defined in the contract
Ideally you want measurement metrics to be defined in the contract (especially when it comes to those software publishers, outsourcers, and suppliers who have

a tendency to audit and magic-up financial penalties). Defining measurement metrics in the contract quashes dubious behaviour around audits or sales that might try to exploit ambiguity for commercial gain. If everyone knows how consumption will be measured, ambiguity can be considerably reduced.

3. Be proactive

You don't want to wait until you're at the end of a three-year contract and renewal process, for example, before looking at your ITAM process. It needs to be a continual, proactive exercise.

4. Collect data automatically where possible

Finally, see if you can collect accurate data automatically where possible, to help measure consumption against licensing agreements. Measurement of consumption should ideally be baked into operations rather than considered an overhead and additional administration task. In this chapter we'll explore some techniques to achieve this.

WHY IS LICENSING GETTING MORE COMPLEX?

To understand why licensing is getting more complex, we need to understand a few of the trends that are shaping modern IT departments. The first trend, and probably the most prevalent, is that of Moore's law. Moore's law predicts that processing power, as measured by the number of transistors, will double every 18 months.

Now imagine a software publisher such as Microsoft selling SQL Server. It is potentially selling a licence to a customer for a flat value, yet the customer is benefitting from more and more powerful hardware supporting the same SQL database. There has been a shift to more power-based or configuration-based licence types to capture the increasing capacity and momentum of Moore's law.

Another trend is that organisations are acquiring more device types. Also consider the huge proliferation of mobile devices and apps installed on them; the concept of personal use and corporate use of devices; and the cloud and virtualisation.

As a result, licensing has become a lot more complex. We've seen a shift to user-based licensing models (since individuals are accessing software from multiple devices) and to power-based models and configuration-based models. It's no longer about counting computers, which is how ITAM began.

A CASE STUDY IN COMPLEX LICENSING

Let me provide an example of how software licensing has become more complex due to innovation. This example requires us to go right back to basics—back to when I first started in the ITAM sector.

It's the year 2000. An IT manager has five laptops with Microsoft Office installed on them; three of them have Standard Edition and two have Professional. If I can demonstrate that somebody with Professional is not using Access (which at the time was the main difference between Standard and Pro), then I could shift that user to a Standard licence and therefore save the company money, not to mention reduce licensing risk, audit risk, etc.

Now consider the modern-day version of Office 365. On the Office 365 website, Microsoft attempts to guide you in terms of the features you might need, in order to direct you to the most appropriate plan.

Options include:

- Do you need desktop apps or just SaaS?
- Do you need apps for Mobile?
- Do you require email as well as Office?
- Do you need file sharing and storage?
- Do you need Skype and HD video conferencing?

It's clearly no longer just about Access!

And that's just the tip of the iceberg in terms of complexity. As you can see, there's so much more to it than counting computers. We need to gather a lot more data types in order to manage Office 365.

For example, when embarking on Office 365, an organisation would need to think about its principles for addressing complex licensing: How are we going to measure the usage of all these elements to ensure we're getting best value? Is it important that we measure it? Will Microsoft catch us out during an audit if we're not able to show how we're managing these features? All these elements should be considered.

"THE RISING IMPORTANCE OF USER MANAGEMENT": FILIPA PRESTON, SOS

Now I'd like to introduce you a practitioner leading the field in terms of software asset management. This is Filipa Preston, the CEO of an independent SAM practice called Software Optimisation Services, based in Perth, Australia. Filipa picked up Microsoft SAM Partner of the Year at the Microsoft Partner Awards in 2016. She kindly shares her views on the importance of user management in modern-day SAM.

Watch the YouTube video of Filipa here:

https://www.itassetmanagement.net/2016/09/08/user-management-rising-importance/

Key points:
- It's possible to do cost control and efficiency and optimisation via user allocation and actively managing users.
- We've gone from counting computers to counting users. It's no good just counting users in isolation. You need to be able to manage the dependencies between users and computers and the relationships there. This is why dependencies are so important.
- User profiles should be a priority for ITAM. If you're not managing users, you face being overcharged.
- If you don't know what your users are up to and whether and their user profiles are up to date, you've got considerable security risks.

- There is cross-vendor benefit to proactive user management. If you actively manage your user base, that's going to benefit Microsoft, but it's also going to benefit other user-centric licence models, such as SAP.

USER OPTIMISATION

Filipa mentions that a risk of being overcharged exists when users aren't managed; I'd like to touch on how to approach optimising user-based licensing because it's becoming such an important licensing model. The optimisation models exist equally for user models as they do for traditional perpetual-licensing models—but they require a new way of thinking.

If you're approaching Microsoft for exploring Office 365, your account manager is going to want to stick you on one simple program and give you the best edition across your entire estate. And why wouldn't the Microsoft sales rep try and sell the top-of-the range edition?

But when you assess the requirements of your user base, you might establish that you have multiple types of users (as per the diagram below). One set of users might need the full edition, but a second set might not require the full sweep of capabilities that the top edition of Office 365 offers. You might even realise that there's a third group that needs even fewer, and then perhaps there's a group that doesn't need Office 365 at all. Maybe they could get by with Google.

The goal, the new role of ITAM in this, in terms of users, is optimising and cutting out all that waste to make sure that we're closely aligned with what the business needs without overspending.

DEPENDENCIES

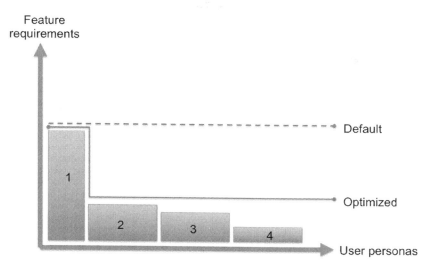

Stella's progress

Stella's company is considering Office 365; it is quite a complex licensing program, with many options for hybrid environments (both on-premise and cloud). How does Stella approach it in terms of the data and dependencies required to manage consumption?

Lisa, the licensing specialist in Stella's team, has been exploring how the company can get the most bang for its buck when it adopts Office 365. She's looking into how kiosk licensing might help save them money.

The "E" range (E1 to E5) is a set of Office 365 subscriptions with different levels of functionality. At the time of writing, each subscription level has an underlying licence requirement. So, Lisa needs an underlying licence for certain subscriptions—for example, she might need a Windows CAL or Enterprise CAL.

A kiosk licence is a type of Office 365 subscription without an underlying licence requirement, so it could generate big savings compared to buying a full E-range subscription for everyone in the company.

Lisa has looked at user profiles, and at her users across the estate. She's identified some people who might be suitable targets for a kiosk licence: truck drivers, factory operators, cleaners, external staff, contractors, etc. She's also taken our advice from the beginning of this module and considered how to measure this.

In terms of metrics, she's can measure users not making use of Active Directory—this is a way of measuring a kiosk licence. Also, she can refer to her company's configuration management database (CMDB) and see if a record of a user without an assigned asset exists. (We'll talk more about CMDB later in this chapter).

> This approach won't suit every company, but the point is that Lisa has a methodology for measuring compliance and consumption that Microsoft has agreed to. It's a way of saying "This is how we're going to measure our position and this how we're going to measure it on an ongoing basis" without the fear, uncertainty, and doubt involved in complex licensing and uncertain audit behaviour.
>
> Lisa has to be aware of certain kiosk licensing rules. For example, kiosk users have to access Office 365 via phone, tablets, or their own computers—it can't be accessed through a corporate device; they don't get Skype or online storage; they can't log into the corporate network; etc. As long as Lisa has got data that justifies how they are managing their consumption, she a great opportunity to shave out many costs from her user base. Especially if she has a lot of blue-collar or factory workers who don't need full access to Office 365.

CMDB AND ITAM

We covered ITAM and ITSM integration in Chapter 8 – Transition. Here, we look at how we enrich the service desk using ITAM data, and in return, how service desk operations automatically update ITAM data. A good example of this ITSM integration with CMDB.

> "A configuration management database (CMDB) is a repository that acts as a data warehouse for information technology (IT) installations. It holds data relating to a collection of IT assets (commonly referred to as configuration items (CI)), as well as to descriptive relationships between such assets. When populated, the repository provides a means of understanding:
>
> the composition of critical assets such as information systems
>
> the upstream sources or dependencies of assets
>
> the downstream targets of assets"
>
> ~ Wikipedia

This is all very wordy—what does this mean in layman's terms? If I receive an alert to say a server has gone down, I can use information provided in the CMDB to determine,

"OK the server that has failed used to host several virtual machines. I know the relationship between the underlying hardware that has failed and the identity of the virtual machines being hosted. Within one of those virtual machines, I know an important business application is running, and it supports sales reps generate revenue for the company." By using the CMDB, I can see the relationship between assets, services, and underlying hardware—I can see the business impact of a hardware failure.

Many service desks are using a CMDB anyway, so this is a great potential source of data for us in the ITAM department. Tap into some of this knowledge to help maintain and watch your licence agreements.

The only caveat is that a CMDB is focused on tracking the relationships between IT services and the configuration items (CIs) that support that service, the CIs and relationships that are tracked with the CMDB generally focus on providing information regarding availability, uptime, and supporting change management understand the risks associated with change and how to manage them. It's not necessarily focused on ITAM goals and the scope of CIs tracked may differ from the scope of the ITAM team. Whilst the CMDB and ITAM complement each other, they don't fit perfectly together. It's rather like a local village hall which might be a nursery during the day and a seventies disco in the evening; they might share the same infrastructure but have very different motives. The CMDB can share data with and even the same database as ITAM, but the two have different goals and priorities.

For ITAM, the things that might be useful in a CMDB are the assets, how they are configured, the dependencies between them, and which assets support which IT services across the enterprise

For example, from an ITAM perspective, we see a SQL Server that needs a licence. But by viewing CMDB information, we learn what purpose the server provides, who the system owner is, whether it is on a stand-alone box or hosted in a cluster, and so on. This is important information when conducting our investigations and also lends considerable credibility to our conversations with other IT department colleagues.

Another caveat when considering a CMDB is that some users and organisations consider it a bit of a unicorn—as in, they've never seen one in real life. Many people can talk a good game about the CMDB but don't have sufficient maturity to actually deliver it, so its data might not quite be suitable, or it might be limited to a very specific scope of devices or services.

In this chapter we have discovered how the complexity of modern IT environments means collecting lots of different data sources to measure our consumption and demonstrate compliance. In the next chapter we'll explore one of the biggest sources of potential savings for an ITAM program, reclaiming assets no longer in use.

Chapter 11

Reclaim

INTRODUCTION

Chapter 9 – Request was about automating the process of issuing new assets to employees. This chapter is about the reverse: getting them back. We'll walk through how to enforce an ITAM reclaim policy to claw back assets whilst also delighting, or at least not aggravating, customers.

As well as reducing the risk of software audits or losing data from missing hardware, reclaim offers the biggest financial returns of the whole ITAM function when executed properly. I'll go through examples of how you might reclaim assets and track usage, discuss the etiquette involved when asking for things back, and outline how reclaim can help you bootstrap your entire ITAM practice if you're struggling to grow as a result of lack of funds.

Throughout this book we've considered the importance of gaining authority and building a solid business plan. But sometimes we need to do a bit of forensic work and stealth research under the radar to build up evidence for doing ITAM. Reclaim is the perfect way to collect financially actionable data to build up your reputation. We'll discuss exactly how you might get started.

WHY DO RECLAIM?

Imagine there's an old company laptop sat in front of you on your desk. It's switched off and collecting dust, surplus to requirements in your company. What opportunities for us in ITAM does this solitary laptop present?

The first thing you observe about the dusty old laptop is that it's an asset. It might not be worth a great deal, but perhaps we can reuse it and extend its useful life—this is the value of asset management. Even if the laptop itself is of no use, perhaps the component parts will be useful to us, or to somebody else.

We can then look at what's installed on the laptop. What software could we possibly reclaim? And remember, software can be reclaimed from a device even if the device is dead. If software was purchased for the laptop and can be legally transferred to someone else—and it can be shown that the original was removed or destroyed—we've saved some money. The basic principle is this: "If I can claw back some software from this old laptop and stick it in a stockpile, I'll save myself money the next time someone requests that software because we won't have to buy it." In most companies who have been buying new IT equipment for decades, there is typically enormous potential for reclaim: recycling or reuse of IT hardware and software.

If you're removing software from the old laptop, you might also be removing the need for maintenance contracts or support renewals against that software. This laptop might also be leased, or have some other service costs against it, all of which could be removed or reduced by repurposing the asset.

A laptop switched off and stuck in a drawer for six months is likely to be un-patched and un-secure within a corporate environment—a potential security threat. And because it's switched off, it's likely we're not meeting our compliance goals because we don't have visibility of it. The existence of this laptop might also be creating false demands or scopes for projects. For example, it might be included in the plans for a new application being installed, or an agreement might be being negotiated based on the number of laptops in the company; if the decision makers knew it was dormant, they might have had different plans. This is the equivalent of ordering pizza for a room of people but without asking if any of them are hungry. By keeping accurate records and marking this laptop as dormant, we're keeping an accurate record of real demand for everyone in the business.

Other overheads that occur for the business as a result of this laptop's existing include the support team's maintaining knowledge of it (just in case it fails); the storage or networking provision made for it when it might not be in use; sensitive customer data, blueprints, or patient records on the device—the list goes on.

This just skims the surface. There are many compelling reasons why you want this asset back in the IT department and out from under people's desks. Some people describe these dormant assets as zombies. They are out of touch and out of control, no longer contactable—we in ITAM really want to bring these zombies back to life, back to being live assets on the network.

The ROI of reclaim

Hopefully you can see the value of reclaiming assets (software and hardware) not in use from an administrative and housekeeping point of view. But about the financial value? What is the potential ROI for reclaim?

Let's look at Stella's team and review what Petra, the inventory analyst, has been doing around reclaim—she provides an example of how we can deliver a lucrative, fast, and relatively easy return for a reclaim project.

Stella's company is going through an operating system refresh. As part of this process, they've identified that they need 500 new laptops. Petra is thinking, "How can we squeeze better value out of our existing assets? Can we look at our existing stock?"

Petra has a budget of $1,000 per new laptop (the cost includes having it prepared and delivered onsite, ready to be productive). Every reader will have wildly different hardware costs, but let's use this as a simple example. Through some proactive ITAM work,

> Petra is able to reclaim 250 laptops from her environment in a short space of time and therefore save the company $250,000 in the new operating system refresh budget.
> Petra saved the company money using a simple method best described as ITAM marketing. It doesn't require any technology—just some hard work and communications.
> She launched a hardware amnesty, stating: "Please give us back your laptops and there'll be no questions asked, no repercussions. Simply hand them back to IT and we'll process them for you."
> This was a simple opportunity for staff to clean out their desks and cupboards or bring in stuff they'd been hoarded at home that was no longer in use. She included the amnesty in the company's email newsletter and intranet site, and stuck up a few posters reminding people of the amnesty, how environmentally friendly the initiative was, and how much money it could save the company.
> This is proactive ITAM: marketing and communications. No process has been initiated or technology deployed. You, as an asset manager, just need to market the idea.

CASE STUDY: "GLAXOSMITHKLINE SHEDS 6 TONS OF E-WASTE"

GlaxoSmithKline is a global company, but this case study refers to one particular office with 1,200 employees spread across 32 floors. The ITAM team shared a notice on the corporate intranet page and through emails and posters. The notice was about the environmental impact of reclaiming and recycling unused equipment. They recruited 35 volunteers, and at minimal cost, they managed to collect six tons of e-waste.

Watch the video here:

> https://www.itassetmanagement.net/2010/02/23/glaxosmithkline-sheds-6-tons-of-e-waste/

They salvaged over a million dollars' worth of hardware (monitors, keyboards, stands, printers and all sorts of other IT junk cluttering up the place) just through that volunteer exercise and a bit of marketing. This is the power of reclaim, even before any tools or processes have been implemented. What could be reclaimed in your environment?

SOFTWARE RECLAIM RESEARCH

Research from the software company 1E suggests that the average software waste ripe for reclaim is $247 per user.[8] The analysis is based on the actual usage data of 149 companies and their 4.6 millions machines, and suggests that 30% of software in the average organisation is unused. How much would that equate to in your organisation?

Key drivers for addressing this waste are:

- cost avoidance (reclaim spare licences and use the stock to fulfil requests, therefore avoiding new licence purchases);
- reduced maintenance (due to lower number of installs) and;
- reduced audit risk (lower software footprint and a stock of surplus software to address/negotiate any shortfalls).

As mentioned, surplus software is also a potential security risk and bloats the network and service desk with unnecessary overhead.

Research from the ITAM Review in 2016 states that "76.4% of organisations admit to being over licensed to defend against audits."[9] This might explain why so much software is unused in the average environment.

RECLAIM BEGINS WITH REQUEST

In Chapter 9 – Request, we discussed automating the request process, building a preferred asset list, steering people's choices, and delivering a great customer

---

8 https://www.1e.com/resource-center/software-usage-report/.

9 https://www.itassetmanagement.net/2016/03/24/business-case/.

experience so that users want to work with the IT department rather than choose other purchasing channels.

Whilst we've discussed ways to claw back software through amnesty, best practice is to build reclaim into the request process: set policy, process, and expectations that when an asset is requested it is issued on a piece of elastic—if you discover it's not in use, it will be twanged back to the IT department.

For example, rather than requesting a copy of Microsoft Project to keep, a user requests a copy of Project to use for six months and then returns it. It's valuable to empower users by giving them access to an app store; let them control their own software. They can uninstall apps not in use or no longer required. Help the users help themselves.

Emphasise that the asset belongs to the company and that if it's not in use, ITAM will find someone else who can use it. It requires a slight change in expectations and thinking: users need to realise they have not requested the purchase of a licence but the usage of software for a period. It requires separating the ideas of entitlement and request.

Think about company cars or trucks in your organisation. If the sales team needs a new car for a new sales rep and a spare one exists in marketing, does marketing cling to it because they had it first? No, an aligned team all pointing in the same direction would recognise that, although marketing made the request, it is a company asset and should be treated as such.

All of these dynamics should be captured within the IT or acceptable-use policy supported by your senior management board, but it's even easier if it's automated in some fashion via an app store.

TRACKING USAGE

We've discussed how you can claw back assets just by asking, but we can accelerate and automate that process by tracking the usage of assets. Over the years ITAM Review readers have shared approaches that cover a whole spectrum; some are dependent on environment and company politics and some are dependent on the type of assets you're managing.

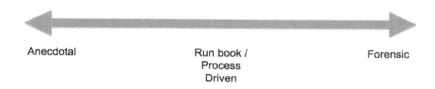

Anecdotal   Run book /   Forensic
                   Process
                   Driven

On the left extreme of our usage tracking spectrum is anecdotal evidence. This could be generated by simply emailing your entire user population and asking, "Is there any software not in use we could remove?" or by sending them a survey. Or you could look at a list of user-based licences and identify if any of the employees had left the company.

At the other extreme of usage tracking is forensic detail to justify reclaim. Tools and technology exist in the market that would record the following: "Martin switched his machine on at 08:30; he went away from his desk for his morning coffee, came back at 08:45, and logged into X application but it was only run in the background—his predominant application was Y; at 10:23 he did 64 keystrokes in application Z..."

There are instances where collecting information this detailed makes a lot of sense. For example, if you're running an expensive engineering application that is priced based on concurrent users, it's sometimes useful to record user behaviour. But for the majority of applications, all we need to know is this: when did the user last use it? Think about the applications in use in your environment and the level of detail you'll need to justify non-use.

There is a happy medium between these two extremes that is process driven, as previously described via the app store examples. If software is recorded as unused for 90 days, for example, then it is automatically removed, or at least a process is initiated to contact the user to see if it's still required.

Ninety days is popular as a time period; software may be valuable even if it's not used all the time, but if something hasn't been used for three months, there's usually a good argument for removing it. If I used an application for submitting my expenses once a month and my company had a thirty-day non-use rule, I'd run the risk of it being removed every month, even though I had a valid business use for it.

Ninety days is also good from a product use rights perspective since some software publishers will allow you to install something and not pay for it if it's not used and is removed within 90 days.

On the other hand, finance departments frequently have software that they use once a year when they prepare the organisation accounts. This software may not be used often, but when it is, it is absolutely business critical as the organisation may incur fines from its regulators if the accounts are not prepared in time.

It's important to engage your business stakeholders to understand how software is used when establishing the appropriate time frames for uninstalling software.

Many of the non-use examples in this chapter refer to traditional software installed on machines. As you build out your reclaim process, you also need to consider modern software in use, for which reclaim could also be implemented. For example, if your IT team is using elastic cloud computing resources such as Amazon or Microsoft Azure, have you got a method of automatically shutting down non-production instances that are not in use? Or for SaaS environments, can you track metrics that will show whether an application is being used or not? For example, Salesforce.com tracks if a sales rep has logged on but also how many new contacts or opportunities have been created.

Finally, from a hardware perspective, whilst forensic usage tracking is available, the most useful metric to track is "have we seen it recently?" Consider day-to-day inventory tracking—if a device suddenly stops responding and being audited, does it mean it's not in use and stuck in someone's drawer? If our asset register said the last person to own it was Fred, let's contact Fred and see if the asset is in use or can be clawed back.

SOFTWARE REMOVAL ETIQUETTE

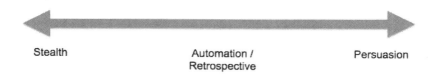

Stealth — Automation / Retrospective — Persuasion

There is a similar spectrum when it comes to software removal etiquette. I don't think there's one definite way of doing this, but I'm going to share with you the extremities in the ITAM market. The approach may depend on the nature of your business.

Some approach software removal as follows: "If somebody isn't using software, I'm just going to remove it without asking them. It's not in use and our policy allows it." This is stealth.

On the other side of the spectrum, people will say, "Sorry, Dear Sir or Madam, Dear Mr Precious customer, I noticed that you're not using this software. Please may I remove it because I'm trying to save the company money? Please, please, please." This is persuasion.

Then a whole spectrum exists between these two extremes. Some vertical markets with thinner margins and more assertive relationships with their end customers, such as retail or construction, might not have much patience and will just remove unused software and apologise later. Contrast that with investment banking, where nothing must be done to upset or irritate the customer; it's a lot more dependent on persuasion. Pick an approach that suits your company and circumstances.

"If they weren't using it, they won't miss it" is a good argument for stealth removal of software without asking permission, but you need to think about the repercussions. Will it create upset users, help-desk calls, and general friction? Or is it a price worth paying for the savings to the company? One way to redress this direct approach is to offer a no-quibble policy. Let's say I've removed 500 copies of Project because they weren't in use. If ten people say, "Hang on, I was using that", you don't question it. You don't argue. You say, "No problem, I'll just put it back."

I've come across arguments whereby a customer says, "Hang on, I was using that" and then the IT asset manager sends them screenshots of the evidence that they're not. I think there's probably a much better way to use your time.

As with usage tracking, a happy medium is to automate the process and involve the user. Notify them that their six-month lease of Project is about to expire and ask if they would like to extend it. If not, it will be automatically removed, with the offer to put it back on if they ever need it.

Remember, you want to empower users to make their own choices; allow them to tidy up their own desktops, their own environments or mobile devices or tablets, if they can. They are much more likely to allow you to take back the software if the request process is easy. We talked about this Chapter 9 – Request; people are less likely to do grey IT, shadow IT, to do IT in the business themselves if ITAM is great to work with, asset request is easy, and things are simple.

BONUS TACTIC: RECLAIM HARDWARE BY STEALTH

I want to touch on some stealth ways of getting hardware back. It's not always easy to prise people's fingers off those laptops stuck in drawers or under desks. If the friendly amnesty approaches aren't working, leverage your friends in the IT security team.

Say you've identified unused laptops in your environment; you've done your amnesty, you've done your newsletters, you've done your posters—but people aren't giving them back. The IT security team doesn't want un-patched devices, potentially unencrypted, with company data sitting dormant in the environment to be potentially switched on at any time. The security team can't track them and meet compliance goals if the devices are never switched on and connected to the network.

So, you need to emphasise that if people hand in old kit, they'll get a better laptop next time they need one. Some artistic licence may be required here! Obviously, you need to be careful what you say—they might not get a better device, per se. But they'll likely get a newer one, or at least a better performing, rebuilt, reimaged machine.

The final and last resort is locking the device. If asking nicely and security requests don't work, lock the device so it can't be logged on to, useful, or connected to the network. Then the laptop is useful only as a doorstop or paperweight. Your asset records can then record the device as archived until it resurfaces. If and when somebody needs to unlock it, it can be re-introduced as an asset to the network. Though this is the last resort, it is very effective.

Asset records

A quick note; if you're proactively reclaiming hardware and investigating usage, make sure you keep accurate ownership records. For example, consider the people on long-term sick, maternity leave, paternity leave, working on oil rigs for six months, etc. We don't want to be pestering them for their laptops. Keep accurate records to be able to declare, "I haven't seen that laptop in inventory for a while, but I know why."

RECLAIM TO BOOTSTRAP AN ITAM PRACTICE

Reclaim can be a great way to build the justification for looking at ITAM in the first place, or to build up some good data to support your business case. On the ITAM Review we often see: "I've got no budget, I've got no authority. What can I do? I do it all by hand in a spreadsheet." Reclaim is a great way to start a snowball of momentum that leads to a fully funded ITAM practice.

You don't need to do reclaim en masse; you don't need to go through a reclaim process for every single hardware device and software install in your environment. Start with one high-end tactical campaign. Autodesk always springs to mind because it's typically high value, sometimes in the thousands of dollars per install, and unused copies can usually be found. It can simply be a case of visiting the architecture or design team that uses Autodesk and stating, "Look, excuse me, I've just got this role, I've got no budget, please help me. Here's a doughnut—please can you tell me whether you're using Autodesk."

Very quickly you can begin to sell ITAM to your senior management team by explaining "Look, I found 20 copies of a $5000 application not being used, which is $100,000. I did that with the clipboard and a few doughnuts and a bit of persuasion. Just imagine what I can do if you give me the budget to do this properly." It's a great way of bootstrapping your entire ITAM practice.

In this chapter we've covered the financially rewarding process of reclaiming unused assets. In the next and final chapter, we'll look at how we verify all areas of our ITAM practice to increase the quality of our results and improve our reputation with verification.

Chapter 12

Verification

INTRODUCTION

We've reached the 12th and final box in the 12 Box framework: verification. In this chapter we will cover what verification means, review concepts in the verification field such as audit readiness and audit prevention, and also refer to the international standard for SAM: ISO/IEC 19770.

Read this chapter to learn how to verify the quality of your ITAM practice and see how important verification is to long-term improvement.

WHAT IS VERIFICATION?

So what is verification and why is it important for an ITAM practice? It's the process of establishing the truth, accuracy, or validity of something. It underpins the credibility and reputation of your output. IT asset managers should have no doubts about the information they are presenting.

For example, if you're presenting reports on risk to your senior management team or stakeholders, it's essential that the information is trustworthy. Any doubts or lack of trust could seriously harm your reputation as a department. Credible results with trustworthy data help us build reputation and momentum within the company. Imagine the scene: you're midway through a big presentation to the senior management team and your IT department peers when someone spots an anomaly in one of your reports. If you are unable to defend the anomaly the tight knit of your pitch might begin to unravel. Verification doesn't make everything perfect, but it demonstrates quality and areas that require improvement so that when you're presenting your findings, you can do so with a degree of confidence.

Verification drives improvement. Through verification, you're checking results, spotting issues and addressing them, and making practices better. Throughout this chapter we'll discover that verification is a real keystone linking and supporting all previous boxes.

We started with winning senior management approval. We took that approval and built a business plan. Then we recruited the team, engaged the stakeholders, and prepared for take-off. We stood up our ITAM practice by looking at entitlement, inventory, and reporting. The practice established, we reviewed major process areas such as transition, request, dependencies, and reclaim. Now, finally, comes verification, which is integral to the whole sequence, the driving force for refining the plan and improving your practice.

AUDIT READY

Digging into the subject of verification in ITAM, we'll trip upon audit defence and audit readiness. We've covered these issues in earlier chapters, but they're worth a recap.

Let's begin with audit readiness. First of all, being audit ready doesn't mean that you have to be 100% compliant at all times. It means that you're ready to manage an audit if one arises. Audits can arise internally or externally, and if you're not prepared they can be time consuming and can suck vital resources from other projects, resulting in un-budgeted spend. Being audit ready means having a plan for when an audit occurs.

For many of you reading this book, the business driver for ITAM is going to be a software audit, i.e., a third-party audit from a software publisher. It's a big driver for many. But it doesn't have to be—audits come in different shapes and sizes. It might be an internal audit. It might be an audit because you're going through due diligence for merger and acquisition activity. It might be a hardware audit for a new lease program. Or it might be that somebody is simply checking the data you're creating. It doesn't have to be only about software audits.

Essentially, audit readiness is about knowing the drill. You have some form of playbook so you can say, "Okay, the audit request is in. I know how to deal with it. I know the data I need to collect. I know how to play this scenario out."

What you want to prevent is headless chickens running about the office panicking because you got a letter from Oracle or Microsoft. Know what the process is.

Stella's plans

In Chapters 1 and 2, we saw that the main focus of Stella's plans was audit defence. There are many valuable reasons for doing ITAM, as we've covered in previous chapters, but for Stella, it's about reducing the cost and risk of external audits.

Stella has begun to build her practice and make some progress, but as she looks to the future, she's looking beyond audit defence and has the long-term goal of taking it to the next level: audit prevention.

In a nutshell, audit defence is reactively defending incoming audit requests as they arrive, to minimise risk; audit prevention is proactively anticipating audit requests and thwarting them from progressing.

VERIFICATION

Audit Defence	Audit Prevention
• Reactive	• Proactive
• Managing Audits	• Thwarting audits
• Collecting evidence	• Known metrics and automation
• Cloud as audit settlement	• Strategic vendor direction
• Stalling & delays to buy time	• Self declaration
• Fire-fighting	• Audit request playbook
• Arguing over audit clauses	• Not being intimidated by an audit

Audit defence is reactive. You respond to an audit request. Audit prevention involves taking the bull by the horns and being proactive. Audit defence is mostly about managing the process. It's about minimising risk and exposure and the cost of any issues that arise as a result of the audit. Audit prevention is about stopping audits from happening in the first place.

In a traditional audit-defence scenario, it's all about collecting evidence, clamouring and scrambling to collect the right data to defend the audit. In an audit prevention scenario, you say, "Hang on. I know what we're supposed to be measuring for that vendor. It's agreed in the contracts, and we've got automated processes in place to collect that data on an ongoing basis. It's not 100% perfect, but it won't be an emergency."

As I write this in 2017, cloud as audit settlement is very much the standard for audit defence. Microsoft, Oracle, SAP, etc.—they want to settle with you for any supposed shortcomings as long as you adopt the latest, shiniest strategic products from them. With audit prevention, we don't need settlements. We can instead focus on what we actually need; we have some strategic vendor direction and we're not just defending and fixing holes using products that we might not necessarily need.

Often the first tactic in audit defence is to delay the audit and buy time. For example, this is a classic manoeuvre: "Right, IBM is requesting an audit. It's with Deloitte. Therefore, we're going to insist on a three-way non-disclosure agreement, and that'll provide the confusion necessary to arrange

> a three-way NDA with legal people across three companies, which will buy us at least a quarter or more to get our house in order and prepare ourselves for the audit."
>
> Audit prevention is saying, "Actually, I know what my position is with IBM because I'm on top of things. I'll do a self-declaration and send it to IBM. I'll do it on my terms and negate the need for an audit in the first place."
>
> Audit defence is about firefighting, about jumping from one audit to the next. Audit prevention is about having a playbook to thwart them.
>
> Finally, arguing over audit clauses is another classic manoeuvre in audit defence: "I'll renew this contract as long as we take out the audit clause" with audit prevention: "I really don't care because my house is in order and I've got the data to defend an audit relatively quickly anyway."
>
> Audit prevention is not a current reality for Stella, but this is what she is aspires to in the longer term. She aims to take the software vendors by the scruff of the neck, to be in control of the audit process so it's not a threat, it's not a risk. This is what we should all be aiming for.

AN INTRODUCTION TO THE ISO SAM STANDARD

The SAM ISO standard is an international independent standard for measuring the quality of your ITAM practice. It has been written by your ITAM industry peers to demonstrate what best practice looks like.

The main best practice international standard for SAM, ISO/IEC 19770-1, is now in its second edition, and several additional components have been published. The major components are as follows:

- ISO/IEC 19770-1 – Best practice
- ISO/IEC 19770-2 – Software ID tags
- ISO/IEC 19770-3 – Entitlement tags

At the time of writing further dashes are under development.

ISO/IEC 19770 CERTIFICATION

There is currently no formally recognised way of being certified against the ISO SAM standard, although a few organisations have asked their SAM partner to assess them against the "spirit" of the standard (to assess their maturity and identify opportunities for improvement).

The ISO standard is divided into four tiers. In theory, an organisation could be benchmarked against these tiers in isolation.

- Tier 1 – Trustworthy data: the ability to generate good-quality SAM data
- Tier 2 – Practical management: having controls in place to manage SAM life cycles
- Tier 3 – Operational integration: driving efficiency
- Tier 4 – Full ISO/IEC SAM conformance: full conformance to the standard

The ITAM Review's free online maturity assessment suggests organisations have either poor, partial, or good coverage in terms of the 12 Box Model's competency areas. If your organisation is hitting well in most areas, it might be worth exploring the ISO standard. Either way, I recommend buying a copy of the standard.

ISO 19770 was written as a SAM standard, but the nature of modern SAM means that in order to do SAM properly, you also need to do hardware asset management comprehensively, so it should be considered an ITAM standard. I expect the name to be changed in the future to reflect this.

Further resources:
- The home for ISO 19770: www.ISO19770.org
- The ITAM Review news archive on ISO: https://www.itassetmanagement.net/category/isoiec-19770/
- A quick guide to ISO -3 tags: https://www.itassetmanagement.net/2016/11/11/isoiec-197703/

HOW THE 12 BOX MODEL MAPS TO ISO/IEC 19770

The table below illustrates how the ITAM Review's 12 Box Model compares to the international standard for SAM (ISO/IEC 19770-1 (2012)).

The major competency areas for the 12 Box Model are in the far-left column, and the major process areas for ISO/IEC 19770 are along the top row.

SAM Process Area	Organizational Management Processes for SAM		Core SAM Processes			Primary Process Interfaces for SAM
	4.2 Control Environment for SAM	4.3 Planning and implementation Processes for SAM	4.4 Inventory Processes for SAM	4.5 Verification and Compliance Processes for SAM	4.6 Operational Processes and Interfaces for SAM	4.7 Lifecycle Process Interfaces for SAM
Topics	Governance, Roles and Responsibilities, Policies/Procedures etc.	Planning, Implementation, Review, CSI	Asset identification, inventory management, asset control.	Asset record verification, license compliance, asset security compliance, verification	Relationship and contracts management, Financial Management, Service Level Management, Security	Change, Software Development, Release, Acquisition, Deployment, Incident, Problem, Retirement
People						
1 Authority	✔	✔				
2 Plan	✔	✔			✔	
3 Team	✔	✔				
4 Stakeholders	✔	✔				✔
Process						
5 Transition						✔
6 Request						✔
7 Dependencies						✔
8 Reclaim						✔
Technology						
9 Inventory			✔	✔		
10 Entitlement				✔	✔	
11 Reporting	✔	✔	✔	✔	✔	✔
12 Verification	✔		✔	✔		

VERIFICATION FREQUENCY

The ISO/IEC 19770 standard suggests that software records be verified quarterly and hardware bi-annually. As mentioned throughout this book, it's all about managing risk and your attitude to it. How often you verify your records will depend on your resources, bandwidth of staff, and your attitude to risk, but the frequency ISO suggests seems quite sensible. Also, consider priorities. For example, even if you have 4,000 vendors installed in your environment, you're only tracking the top ten.

When it comes to verification, you also need to consider building automation and processes. You don't want to end up in a position where you're saying, "Oh right. I've got to go through my verification process because I haven't yet this quarter." You should build in steps that do verification for you automatically. I will touch on those shortly.

The final thing to think about in terms of verification frequency is the reports and dashboards you're presenting to your management team and stakeholders. These are the key things to verify because these will be what gets scrutinised. For example, if you present a Microsoft ELP to your senior management team to say "This is our position at Microsoft because it's a massive vendor for us", you need to think about the underlying data you'll be quizzed on—that's what needs to be verified.

Verification should be your number one priority. If you're in a management meeting and they start digging into this data and the data hasn't been verified, you'll end up with egg on your face, embarrassed in front of your peers, and people will lack confidence in your practice.

Stella's verification plan

Stella needs to return to the very first box: authority (i.e., senior management approval for her business plan). She set out a vision acutely aligned with what the business was going through. It said, as I've repeated a number of times throughout this book,

"The ITAM department will provide inventory of all assets with 95% accuracy, and an audit-ready status will be maintained for the top ten strategic software publishers with 97% accuracy."

This is a very focused message, and I will repeat myself once more: this is not necessarily the message that's right for you. You need to think about what's relevant for your business. If necessary, go back to Chapter 1 to decide what that might be.

The verification process is already built into Stella's vision. In order to justify her business plan, to justify that she's doing the right thing, she needs to report on the ELP for her top ten strategic suppliers and the trends that are happening in this area. She also needs to report on inventory and visibility and the trends happening there, and she needs to produce a log of audit activity. In short, any verification that she does needs to be focused on these three areas.

Verifying inventory

We covered how to measure accuracy for inventory and how to verify inventory data in Chapter 6 – Inventory, but here's a quick recap.

We met with Petra, the ITAM analyst working for Stella, and she examined the previous audits that occurred within her company. She identified that the auditors were looking for two main things in terms of verification: have you accounted for everything and have you got complete visibility of your environment?

Petra decided that in order to verify her inventory data, she would do three things.

She did physical spot checks, which basically means approaching a colleague, preferably outside the IT department, and actually checking their device: "Right, for this laptop or this server or this mobile device, is what's being reported entered in the inventory system or my ITAM management repository? Does it accurately reflect what this device has got on it?" If not, then you've got some serious issues. Doing spot checks for a small sample of your environment is a way to validate your work and provide you with the confidence that things are working correctly.

Next, she did life cycle checks. For example, say a laptop gets returned to the service desk to be repaired. As part of the process, the service desk analyst should look at the inventory record and compare it to what's on the laptop—this is part of the asset's life cycle. Is inventory accurately reflected with what's on this device?

Finally, Petra she did a comparison with other data sources. She took her inventory data and compared it against Microsoft SCCM data. Then she compared it against Active Directory records. Her goal was to spot devices that were missing to drive accuracy and therefore trust in her data.

"THE ROLE OF VERIFICATION IN ITAM": RORY CANAVAN, SAM CHARTER

Rory is an independent SAM expert, long-term supporter of The ITAM Review and all-round good egg. In the video below I asked Rory about his view on verification of ITAM data.

View the YouTube video of Rory here:

https://www.itassetmanagement.net/2016/12/07/role-verification-itam/

Key points:
- The reason for comparing data sets is to drive accuracy; any blind spots might trip you up at audit time.
- There is nothing worse than a senior manager's digging into your reports and spotting inaccurate data.
- Its not just about spotting exceptions but also about remediation of issues and root cause analysis to identify how issues occurred in the first place.

Entitlement verification

We've covered verifying inventory, but we also need to think about verifying entitlement. Back in Chapter Five – Entitlement, we walked through a scenario of fully exploiting product use rights to reduce compliance exposure. We met with Lisa, Stella's licensing specialist, who used product use rights and unused software to reduce an exposure of $1 million to a $50,000 overage by using SAM and entitlement techniques and squeezing best value out of entitlement.

Lisa needs to ensure that the data behind her arguments is solid. She

> should be asking herself: "Is it still the case that you've got 1000 licences of Factory Standard? Has this been verified?" And, "Is the inventory record of 2,000 accurate? How old is that data? We've got downgrade rights for Factory Professional that we're applying, in order to achieve significant savings. Are those downgrade rights still applicable? Have we engaged in some sort of agreement that wipes those out?" Also, Lisa sees unused software: 40% of the applications installed for Standard are not used. "Is this accurate? Have we validated the usage data to see if we can actually remove them?"
>
> Lisa needs to think about the data (e.g., the ELP) she's presenting to management teams. She needs to verify it to drive accuracy. It will be a benefit to Lisa and also add significant weight to her reports.
>
> At the beginning of this book we looked at how Stella planned to maintain momentum in her ITAM practice. She needed to get the CIO's blessing, take that forward to a monthly or quarterly board, and then drive through a continuous service improvement plan. Verification is very much a part of that.
>
> Verification demonstrates that not only is Stella making progress, but also, her data is credible. Organisations with some form of review process in place, like a monthly board, alongside a verification process, are in a strong position to improve and build a world-class ITAM practice.

In this chapter we have explored how verification can add credibility and quality to our ITAM practice, ensuring our information is accurate and ready to satisfy audit level requirements.

What next?

Thank you for reading my Practical ITAM book. I hope it has helped you begin to understand the world of ITAM and will support you in making an impact with your ITAM practice.

My goal was to help you understand what your top priorities are, what should be done first, and how to build a valuable and lasting ITAM practice for your company.

RECOMMENDED APPROACH

As mentioned in the introduction, once you have familiarized yourself with the key principles of the 12-box model you can take our free online maturity assessment to identify the strengths and weaknesses in your company (The maturity assessment is available via the ITAM Review website).

Please also take advantage of The ITAM Review community to ask any questions you might have and get support from your peers. Finally, if you wish you can verify your learning by taking the 12 Box certification exam.

Good luck and let me know how you get on!
Cheers, Martin.
https://www.itassetmanagement.net/practical-itam/

The Author

I started working in the ITAM field in 2000, when I joined Computer Associates (CA) in the channel inside sales team. There were about ten of us in the team, and most were focused on CA's big earners at the time—Arcserve Storage and eTrust Security. It was my role to look after the smaller product lines that existed outside of these big hitters; I focused on things like software distribution, remote control, and this rabbit hole of a subject called ITAM. The rest of the team affectionately referred to my product lines as "weird-ware".

As with most people who stumble into their career in their early 20s, I didn't choose my field—ITAM chose me. But I've been proud to specialise in the subject and be part of the global community ever since. After CA I worked at ITAM-based consultancies and software companies, including Centennial Software and FrontRange Solutions.

I founded the ITAM Review in 2008 so that anyone involved in the SAM or ITAM industry could share their expertise, feedback, and opinions for the benefit of others.

I'm proud to provide an independent platform for worldwide professionals to share their practical implementation experience among peers.

Today, the ITAM Review hosts conferences worldwide so that IT asset managers from across the globe can connect and learn. The ITAM Review also provides education and independently verifies tools and service providers. The community attracts over 1.3 million visitors a year, including ITAM, SAM, and software licensing professionals around the globe.

On a voluntary basis I'm the chief agitator at the Campaign for Clear Licensing, a contributor to ISO WG21, which develops the SAM International Standard ISO/IEC 19770, and a board member of Free ICT Europe, which supports the ICT secondary market.

www.clearlicensing.org
www.19770.org
www.free-ict-europe.eu/

I have three children and live in Swindon, Wiltshire, in the United Kingdom. When not working I can usually be found spending time with my young family, travelling, or exploring the Wiltshire countryside on my motorbike.

The ITAM Review Newsletter

Subscribe to the free ITAM Review newsletter and receive:

- Monthly industry updates delivered straight to your inbox
- Discounts and special offers
- Links to training and useful resources
- Access to expertise, feedback, and opinions from industry experts
- News about our upcoming events

http://www.itassetmanagement.net/contact/subscribe/

GET INVOLVED: JOIN THE ITAM REVIEW COMMUNITY

At the ITAM Review we believe everyone in the ITAM market should be heard—from the one-man band consultant to the largest software companies in the world. The ITAM Review wants to hear from all ITAM practitioners from all countries and backgrounds; we want to receive updates from the newest of start-ups to the oldest of industry stalwarts.

Contributions are free and most welcome. It's a simple trade; your expertise in exchange for raising your profile among ITAM Review readers. The only thing we ask is that it's not a sales pitch for your products and services—let your expertise do the talking!

LET'S TALK!

To contact me, please use the links below:

- LinkedIn: http://www.linkedin.com/in/martinthompson
- Contact page: http://www.itassetmanagement.net/contact/

Made in the USA
Lexington, KY
27 January 2018